Agile Hardware Product Realization

Mastering the Journey from Concept to Scale

Michael Keer

Founder and Managing Partner,

Product Realization Group

Illustrations and co-author:

David Eden, COO, Product Realization Group

Agile Hardware Product Realization.

Version 2.1

Copyright © 2023 Michael Keer

ISBN: 979-8-38781-133-3

All trademarks are trademarks of their respective owners. Rather than put a trademark after every occurrence, we use names in an editorial fashion only and to the benefit of the trademark owner, with no intention of infringement of the trademark. Where such designations appear in this book, they have been printed with initial caps.

This book is a work of nonfiction. Although the author has made every effort to ensure the accuracy and completeness of the information contained in this book, the author assumes no responsibility for errors, inaccuracies, omissions, or any inconsistency herein. Any perceived slights of specific people or organizations are unintentional.

This work is provided "as is" without any express or implied warranties. While the author has made every effort to provide accurate and up-to-date information, the author makes no representations or warranties of any kind, express or implied, about the completeness, accuracy, reliability, suitability, or availability with respect to the information, products, or services contained in this book for any purpose. The reader assumes full responsibility for the use of the information provided herein and agrees that the author is not responsible or liable for any claim, loss, or damage resulting from its use.

"This book deeply resonates with my experience. The NPDI processes contained within are of tremendous value to hardware development teams. It is a must read for design and manufacturing professionals!"

Robert Bisaillon, VP Operations, Synopsys Inc.

"Michael's insights, methods, and tools provide for an extensive roadmap to successful product realization. From resource management to MVP and beyond, Agile Hardware Product Realization packs decades of experience and understanding of the product development process into easy understandable methods and processes that any aspiring startup team can utilize in a straightforward manner. In particular the areas of scale, manufacturing, and supply chain take the reader through some of the most important areas that are usually neglected or over-simplified by leaders who focus too much on the design and specs of a new product. I recommend a thorough reading of this book and its principles by any team regardless of its experience as a key step to help assure the success of their new venture."

Martin Lynch, Chief Operating Officer, Freewire Technologies

"It's a well-known adage that launching a new hardware product is hard, but this book will help smooth the way. With a clearly defined process distilled from industry veteran's best practices, it offers a guide to help you navigate new projects from concept to launch. As a testament to its effectiveness, we followed this process to successfully launch Roost's first product in under a year, thanks to the early identification of hidden roadblocks. The invaluable lessons in this book are now deeply ingrained in our product development process, paving the way for another 10 successful product launches and beyond."

James Blackwell, CTO & Cofounder, Roost Inc.

I dedicate this book to my father, Leon M. Keer. A loving father and devoted university professor, he has been a source of inspiration for both me and the countless others whose lives he interacted with throughout his lifetime. His spirit lives on through this work, and I hope to honor his memory by sharing the knowledge he instilled in me.

Michael Keer, April 2023

Foreword

In November of 1805, the famed explorers Lewis and Clark arrived at the western edge of North America and looked out over the Pacific Ocean. Their journey had taken almost a year and a half and was filled with stories they captured in their detailed diaries and hand drawn maps. Many Americans waited with anticipation for Lewis and Clark's return - hoping to read the accounts of their expedition, and to imagine taking the journey themselves.

Their journals and maps were published in 1814 and quickly sold out. Adventurous traders and optimistic settlers bought the books hoping they could retrace Lewis and Clarks route to find a new life. Today, these rare first edition journals can be found in private collections and university libraries around the world. But one thing that's even more rare, is an original edition still containing the fold out maps. Anyone that attempted the journey didn't pack the heavy book, instead they cut out the map and folded it into their pocket. For them, it was the most valuable part of the book.

Today, the western edge of the United States is filled with established tech companies and countless startups, hoping to develop products that will change the world. And in every case, there's an adventure that's taken place as that product was brought to market. A daring journey from the first optimistic idea to the product release. And sadly, many products never make it into the market. Michael Keer has carefully mapped and journaled his countless routes along the New Product Introduction (NPI) process. And he's provided it to each of us, so that we might prepare appropriately for our own adventures.

My hope for you, as you read this book, is that you will find yourself embarking on the adventure of a lifetime. I did just that 7 years ago, and Michael was my guide for much of the journey. If your NPI adventure is like mine, it will be filled with some ups, some downs, moments of contemplation and lots of realizations. And ultimately, if well plotted, your expedition will be filled with much success as you look out over the market and observe the fruition of your efforts.

So, with that I'll give you some advice: Dream big. Work hard. Imagine success. And most importantly - don't start your journey without the map.

Mark Frederick,

Chief Technology Officer and Co-founder, MākuSafe Corp.

Contents

List of figures

Preface

After $300 hundred million dollars and three years into my first Silicon Valley startup, Raynet Corporation, this innovative optical telecommunications startup got hit with a serious blow from Bellcore, the telecommunications regulatory body, who rejected its "fiber in the loop" architecture that was the basis of its competitive advantage. Unfortunately, this meant that the product, as designed, could not be shipped to customers.

To fully comply with Bellcore, it took three more years and an additional $300 million dollars. Making matters worse, this new architecture eliminated the product's fiber bending competitive advantage. After the new product was revised to comply with Bellcore's requirements, growing sales were not forthcoming, and the primary investors became increasingly concerned. Sadly, after more than six years of development and $600 million in funding, the business was sold to Ericsson for roughly $30 million – a catastrophic return on investment. One upside was that over 800 innovative technologists quickly dispersed to companies such as Cisco, Finisar, Ascend Communications, Google, Apple, and many others. These colleagues and friends became strong network connections for future endeavors, fulfilling the Silicon Valley adage that company names may change, but the names of the talented people supporting them remain.

As a young, idealistic engineer, this experience opened my eyes to the unanticipated risks (and costs) that high-tech companies face in bringing innovative new products to market. Several years later, I was Senior Director of Operations at Paramit Corporation, a Silicon Valley electronic manufacturing services provider. Over the eight years I worked there, I saw an increasing gap between our clients' need to ship versus their ability to create, communicate, and manage changes for the sets of documentation needed to build their products at scale. It struck me that few involved in the product realization process seemed to care about bridging this gap, and so I formed the Product Realization Group (PRG) to help companies bridge the gap between concept and full market scale.

The intent of this book is to share the many years of "hands-on" experience and wisdom from the members of PRG in combination with outside industry experts into a book that will increase the awareness of what it really takes to get a hardware-based product to market at scale. By sharing these experiences, wisdom, and expertise, my desire is to demystify the process of bringing new hardware-based products to market in an agile way as well as help businesses avoid common unintentional and costly mistakes.

Because this book spans the entire product realization process, it covers each practice at a fairly high level. If you desire more details about a specific area (e.g., development or supply chain), there are many books available that focus on these areas in detail. The unique value added of this book is its holistic view of the entire product realization process, with practical tips provided for each of the practices described.

We've grouped this knowledge and wisdom into 10 best practices of agile hardware product realization. These practices are geared for those directly involved in the new product introduction process of hardware-based (or system-based) high-tech products. If these practices are implemented properly, they will help reduce your risk for bringing new products to market and increase your chances of success in the market with a byproduct of scalable business processes to support the introduction of future products.

If you are involved in the process, or are just curious about how the process works, then this book is for you. It assumes that you have already done the heavy lifting of marketing discovery to the point where your product definition is clear and documented properly via a Marketing Requirements Document (MRD) and a Product Requirements Document (PRD) as robust inputs to the process. It is worth noting that both of these documents can be considered as "living" throughout the life of the product; that is, they will change as new challenges present themselves or new ways of achieving results are found.

I would like to thank all the seasoned consultants, clients, and dedicated service providers who contribute actively to the hardware product realization process on a daily basis. Many of these contributors fly under the radar of high-profile high-tech startups and established corporations. These "unsung heroes" have dedication and passion for bringing new products to market and are adept at getting things done in the background to help these businesses mitigate risk and enable them to wow the market with innovative new products.

Michael Keer

Acknowledgments

I am incredibly grateful to the brilliant team who has helped bring this book on hardware product realization to life. Their expertise, dedication, and passion for the subject matter have shaped this work into what it is today.

First, I must express my sincere appreciation for my co-author, editor, and outstanding graphic artist, David Eden. His creativity and attention to detail have elevated both the content and visuals throughout this book. My genuine gratitude also goes to our contributing editors, Jane Nevins and David Couzens, whose keen eyes and expert insights have significantly improved the quality of this manuscript.

I would be remiss if I didn't acknowledge the invaluable contributions of our expert reviewers, who generously shared their time and knowledge. A heartfelt thank you to Allen Adolph, Howard Edelman, Jay Feldis, Wayne Firsty, Michael Freier, Shirish Joshi, Ken Kapur, George Lewis, Sunil Maulik, Wayne Miller, Feroze Motafram, Sheila Walsh-Pickering, Jeffery Rosen, and Fred Schenkelberg. Your feedback and wisdom have undoubtedly made this book better.

On a personal note, I must express my deepest gratitude to my wife, Cindy. Her unwavering patience, understanding, and support have made this remarkable journey possible. I am truly blessed to have her as my partner in life.

1 Introduction

By reading this book, you will gain the wisdom from knowledgeable experts that span hundreds of years of experience and substantially increase your ability to meet new product time-to-market goals, lower costs, reduce risk, and set a solid foundation for future growth. This book applies agile hardware development concepts combined with best practices for scaling into volume manufacturing to the entire product realization process.

This book is intended as a practical guide for those engaged in the technology-based hardware product realization process, from conception to full market scale, and for all sizes of business and product complexity, from startups to Fortune 50 companies. An agile and robust New Product Introduction process is important for all hardware companies independent of their size and lifecycle status.

As you explore the book, an important element of your New Product Development and Introduction process to consider is the type of product that you intend to move from concept to volume production. Less complex and higher volume products involve very different development and operational strategies than more complex and lower volume products.

If you are short on time, then you may want to skip to Section 6.1 to help you gain quick insights into the areas where your business is at the greatest risk.

For companies bringing their first product to market, there will be a great opportunity to implement best practices from the start and to "get it right the first time." For companies with established processes and systems, there is an opportunity to revisit, retool, and optimize your existing process. As you likely already know, actively changing established business processes and systems will require strong management support, a desire for cultural change, and companywide dedication to achieve sustainable results.

I recommend reading the whole book to demystify the process and gain the full value offered. By doing this, you will learn about areas you may not typically become involved with and understand the importance of how these practices reinforce each other to help you get better products to market faster with less risk.

2   Product Realization Overview

2.1   Defining Product Realization

Product realization is a term used by the International Organization for Standardization (ISO) 9001:2015 and refers to the basic design and realization of a product provided to customers that are measurable by quality control. It provides clear, certifiable standards for the process of bringing a product to market. Because this book encompasses multiple elements throughout the product lifecycle, you may see terms such as New Product Development (NPD) and New Product Introduction (NPI), as well as New Product Development and Introduction (NPDI). Regardless of the three- or four-letter acronym used, the intent of this book is to share a framework that combines agile hardware development concepts along with best practices for scaling into volume production to increase your chances of success and reduce your risk in bringing new products to market.

2.2   Product Realization Flow

Transforming a product from concept to volume manufacturing requires a journey through a series of stages or phases of activity increasing the details of knowledge about it and reducing the uncertainties. The standard sequence of phases is depicted in Figure 2.1 below.

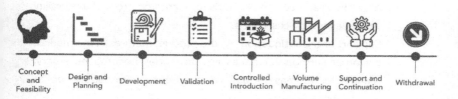

Concept and Feasibility   Design and Planning   Development   Validation   Controlled Introduction   Volume Manufacturing   Support and Continuation   Withdrawal

Figure 2.1: Product Realization Flow

The phases of the product realization flow are described below:

- **Concept and Feasibility**: This initial stage is where ideas for product development are generated, market viability is assessed, technical feasibility is determined, and essential product features are identified.
- **Design and Planning**: In this phase, high-level ideas and concepts for the product are translated into concrete plans, with cost estimations, timeline and resource requirements being established.
- **Development**: This phase encompasses all engineering design and development work, including prototype creation and risk reduction efforts associated with product requirements. Agile hardware development may involve multiple rapid prototyping cycles within this phase.
- **Validation**: This phase involves formal product testing to verify that all the requirements have been met before transferring the design to manufacturing.
- **Controlled Introduction**: During this phase, volume manufacturing set up is performed to ensure cost-effective production. Concurrently, customers are provided with the first complete products to experience, including all the packaging, documentation, training as well as the support and returns processes.
- **Volume Manufacturing**: In this phase, products are manufactured in quantities sufficient to meet sales demand, guided by sales projections and customer needs to guarantee product availability when desired by customers.
- **Support and Continuation**: As the product matures in the market, customer support for issues and upgrades become necessary. Manufacturing may require replacement or upgrades due to unavailability of parts.
- **Withdrawal**: Eventually, the product will need to be withdrawn from the market. This decision may stem from the introduction of a newer model, because low sales numbers have made it unprofitable, or because it is no longer viable to manufacture.

2.3　Phase-Gate Process

As you begin your product realization journey, implementing a phase-gate process (see Figure 2.2) will enable you to assess the viability of the project as it progresses through the product lifecycle.

At the end of each phase, a formal team-based review is held to evaluate the project viability relative to a specified exit criterion. During the review process, the team will assess the progress of the plan, resource allocations, risks, costs, and potential pivots required for the project to move forward. If the project fails the gate review, then it will either be restructured to pass in the future or cancelled altogether (embracing the concept of "fail fast") to avoid increasing sunk costs for a nonviable project. If the project passes successfully through the gate, then it will move forward to the next phase, as shown in Figure 2.2.

This type of phase-gate process may seem to be at odds with the traditional ideal of an agile process, however, but for hardware, owing to the increasing cost of change during the latter phases of this flow, it is important to maintain regular check points on progress. Later sections of this document show how agile repetitions can be built into several of the phases described above as part of agile hardware product realization.

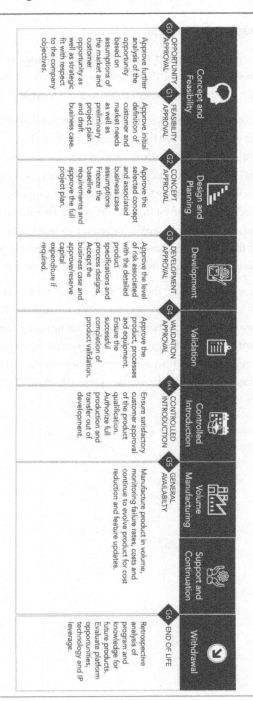

Figure 2.2: Typical phase gate process.

As your product progresses through each key gate in the process, you will want to ensure that all team members are aligned and in agreement before moving to the next gate. Figure 2.3 illustrates, at a high level, some of the key activities and dependencies that are required to support the product realization process.

The combination of multiple disciplines that must be organized and applied in the proper order through each gate along with the need to meet outside regulatory requirements makes developing physical products more expensive, time consuming, and risky than developing pure software application products. Many hardware products fail because of a lack of understanding of what is truly required to navigate the product realization process.

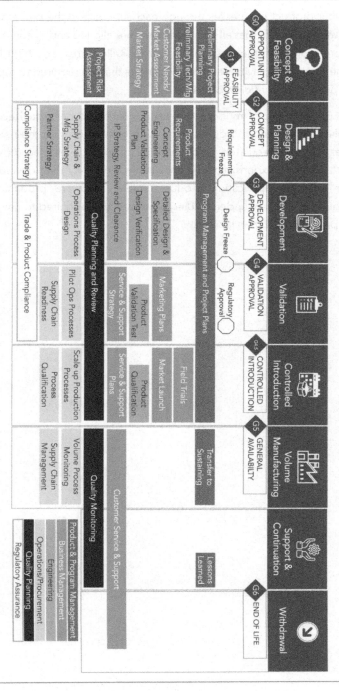

Figure 2.3: Phase gates showing cross-functional dependencies.

3   Understanding Your Markets

Market needs and expectations vary widely, so it is important to decide which market to initially prioritize and then to understand its unique requirements. Every country or region has its own set of rules and regulations, competitive environment, culture, and language, all of which may directly affect the product design. Consequently, where you are shipping to directly affects the products you are developing. Therefore, developing a clear go-to-market strategy that includes a country or region roll-out plan as part of your portfolio management is critical to making the most of your development resources and conserving cash in the early stages of the lifecycle.

3.1   Impact of Shipping on Your Development and Deployment Plans

Thinking about where you want to sell your product before you start development is extremely important and many factors can influence the choices that need to be made, as Figure 3.1 illustrates.

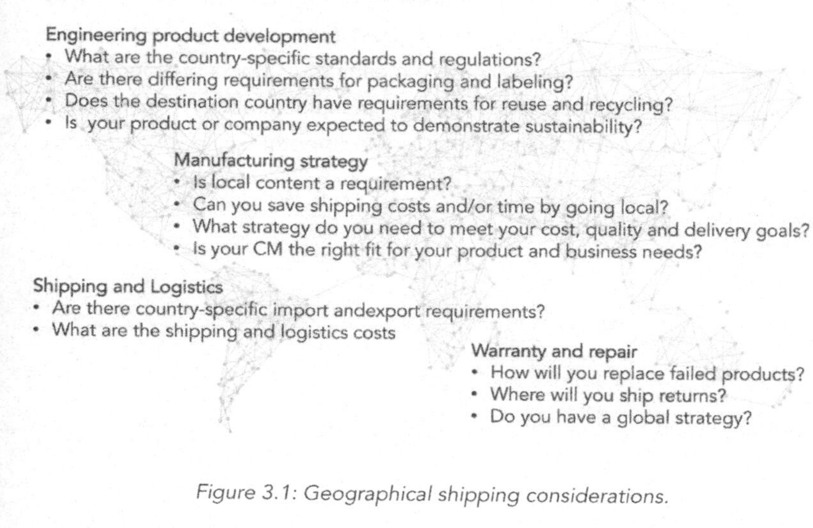

Engineering product development
- What are the country-specific standards and regulations?
- Are there differing requirements for packaging and labeling?
- Does the destination country have requirements for reuse and recycling?
- Is your product or company expected to demonstrate sustainability?

Manufacturing strategy
- Is local content a requirement?
- Can you save shipping costs and/or time by going local?
- What strategy do you need to meet your cost, quality and delivery goals?
- Is your CM the right fit for your product and business needs?

Shipping and Logistics
- Are there country-specific import and export requirements?
- What are the shipping and logistics costs

Warranty and repair
- How will you replace failed products?
- Where will you ship returns?
- Do you have a global strategy?

Figure 3.1: Geographical shipping considerations.

3.2　Aligning Your Business with Your Product Realization Strategy

Depending on where your business is in its lifecycle, your product realization strategies should be tailored to best support your unique product profile, market opportunity, and business dynamics.

The following are some high-level strategies to consider for different stages of the business lifecycle:

- **Early-stage startups**: This is the "garage" phase during which resources are highly constrained and funding is very limited. Your focus here should be on understanding your markets and building your business case, assessing your risks (see the Understand and Mitigate Risks Early section), and developing a functional prototype that can be used to support your next round of funding. Manual and semiautomated processes should be implemented and optimized so that fully automated systems can be smoothly incorporated as the business evolves.

- **Rapidly ramping businesses:** This is your opportunity to set a solid foundation and business systems infrastructure to rapidly scale your business. Time to market will often take precedence over investment. Following the 10 best practices of agile hardware product realization will have a meaningful impact on your time to market, product performance, manufacturability, compliance, reliability, profitability, and ability to smoothly scale your business. Don't try to cut corners here as they will come back to bite you down the road.

- **Established corporations:** NPI (or product renewal and updates) for established business presents an excellent opportunity to review and improve existing legacy development practices and systems and evolve to a more effective, innovative, and efficient agile approach. Start with senior management support and plan the time and resources to shift your culture as well as your systems infrastructure to embrace innovation and operate more nimbly as a team. Over time, you will see substantial improvements in product portfolio management, operational execution, competitive positioning, and the ability to bring new products to market faster with less investment.[1]

---

[1] Note: Many of the techniques listed in the book are supersets of activities that should be evaluated and tailored to best fit your unique product profile and business needs.

4 Distinguishing Agile Hardware from Agile Software

The desire to gain a competitive advantage by releasing products faster has driven the increasing deployment of agile software methodologies. Recently, a push to apply agile techniques to the hardware development process has resulted in varying degrees of success. Because electronic hardware products cannot be fully tested until subassemblies have been physically built, modifications to the product realization process between agile software and agile hardware are necessary. Based upon many years of practical experience and PRG client engagements, we share which methodologies we believe work best for agile hardware-based product development. For products regulated by the U.S. Food and Drug Administration (FDA), there are design control requirements that may make it difficult to eliminate the waterfall approach. A common strategy for these companies is to incorporate agile practices in the concept phase, where less design controls are required, until much of the product development risk has been mitigated and then shift to a waterfall-based approach upon entering the design phase.

In traditional (waterfall) product development processes, design, and testing activities are often performed in sequence, with each phase completed before the next one begins. This approach can lead to long development cycles and the need for costly and time-consuming rework if problems are discovered late in the process.

Agile software methodologies, in contrast, allow for rapid iteration of designs and adapting to changes. Figure 4.1 shows the conceptual differences between the two methodologies.

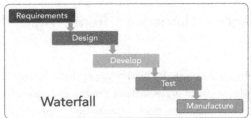 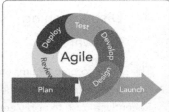

Figure 4.1: Traditional waterfall and agile methodologies (conceptual).

4.1 Key Elements of Agile Software

The Agile Manifesto lists four key elements that support rapid iterations in the software world:

- Value individuals and interactions over processes and tools.
- Value working software over comprehensive documentation.
- Value customer collaboration over contract negotiation.
- Value responding to change over following a plan.

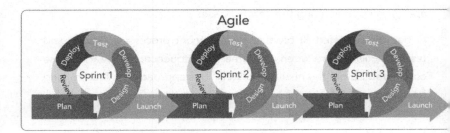

Figure 4.2: Standard Agile sprints.

Figure 4.2 illustrates how the concept of agile "sprints" are implemented in a development environment. For hardware-based product development, the timing of sprints may be elongated.

4.2   Key Elements of Agile Hardware

Through PRG's decades of hands-on experience supporting the development and introduction of hardware-based products, we define the key elements of agile hardware as follows:

- Develop a concurrent product development framework supported by a strong NPI process and software-based business systems.
- Deploy working hardware that includes critical validation testing, design for excellence reviews, and appropriate documentation.
- Engage in active technical program management to identify and mitigate risk.
- Establish customer collaboration along with an in-place preliminary contract.
- Devise iterative product development cycles within the framework of a strong NPI plan. Figure 4.3 illustrates this point.

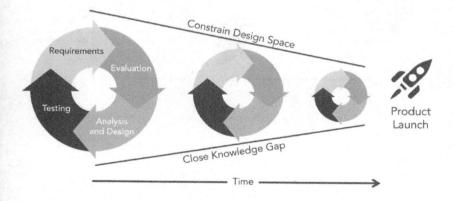

Figure 4.3: Agile hardware iterations.

A well-structured agile hardware development plan will gradually constrain the degrees of freedom to make design changes over time as the project progresses to product launch. Figure 4.3 shows what most people think will be the development path. There is a clear goal and there are iterative cycles of innovation (examining requirements, evaluation, analysis, and design followed by testing against the requirements). The intent is that this approach will produce better and better prototypes on the journey toward full realization into volume manufacturing.

However, there is rarely a smooth straight line of continuous improvement and innovation between concept and scale. Be prepared for this journey to take several twists and turns and expect that the final shippable product may be far from the original envisioned in the initial prototype! Figure 4.4 illustrates this nonlinear journey that most companies experience.

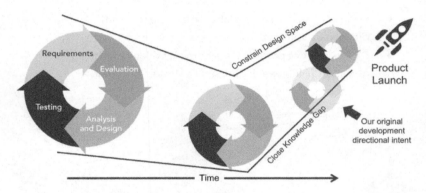

Figure 4.4: Realistic journey of agile hardware iterations.

5 Agile Hardware Product Realization

5.1 Impact of Following the "Bad Old" Ways

Developing any hardware-based product requires planning, management, design, and testing before it can be manufactured and shipped to customers. In the traditional "waterfall" development flow, this work is done in a series of staged activities run in series, one after the other, with review and approval required before passing from one stage to the next. This has merits for certain types of development, for example ones in which an external proof of progress is required to gain further funding. Because activities are done serially, this traditional "waterfall" methodology requires more time to get a product to market, which increases both development cost and market risk.

In any hardware product development, it's not just the technology that is being designed or tested. In fact, many other factors play into the success of a product. Including all the requirements of these characteristics within the finished product necessitates an extensive team of people with diverse skills and strong collaboration between these teams at every stage of the design process.

The need for ongoing communication between all the teams and all of the design considerations can slow each stage of the waterfall process down to a trickle because not all information is available at each stage. As a result, people and processes often must wait for stages to complete before they can move onward to the next phase.

Companies that follow a traditional waterfall and/or ad hoc development process will often experience the product–profit lifecycle curve shown in Figure 5.1.

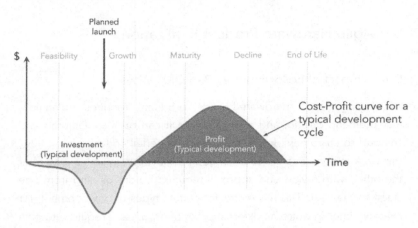

Figure 5.1: Typical product development cost–profit curve.

5.2   Difficulties in Bringing the Right Team Together

The complexity of bringing together the right resources at the right time and then scheduling the project so that these resources are best applied is both a challenging and risky process. The "ideal" project plan (Figure 5.2) for the product often starts with a shipping date in mind and then squeezes all the activities and teams together to meet that future date.

Figure 5.2: Example of the ideal plan.

5.3   Missing Interactions in the "Ideal" Project Plan

Companies often underestimate the time and investment required to bring the right level of skills and the range of disciplines (such as regulatory, reliability, and testing) to successfully launch a new hardware-based product. The "ideal" plan above will often miss many of the low-level interactions and dependencies between teams as well as the competing constraints for internal resources. This occurs because of the need to balance reporting between the intricate detail of the day-to-day work and the major review points such as phase gates (described later in this book). Consequently, these interactions are hidden beneath the weekly or monthly reporting. As a result, the true difficulties and risks of the development are hidden, leading to late delivery and undesirable feature reductions in the final product as the teams struggle to keep up.

5.4   Challenges in Aligning Resources to the Project Schedule

While many companies understand the importance of bringing the right disciplines and expertise together needed for launching a new hardware-based product, these businesses often neglect the ongoing interactions necessary between disciplines, and, as a result, they miss key development and product launch milestones. For example, a company PRG supported for regulatory certification a few years back, after the following issues occurred, neglected to engage their development resources for a pre-submissions assessment during the development phase. As a result, the product failed regulatory compliance testing and was rejected by the regulatory body three different times. To finally meet compliance, the business was forced to redesign and resubmit repeatedly, which caused their product launch to be delayed by six months. If the regulatory and engineering resources had been better aligned to support the project schedule, this six-month delay could have been avoided. Figure 5.3 illustrates how continuous interactions act like glue to keep the NPDI team connected through the NPDI process.

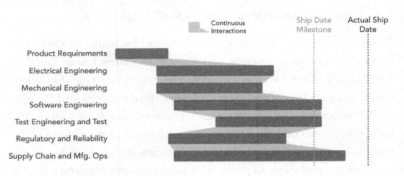

Figure 5.3: Example of the ideal plan including interactions.

Companies that follow this traditional development process typically underestimate the costs, resources required, competing internal priorities, time, and complexity of the product development system that it takes to bring physical products to market. As a result, they will face market launch delays, products missing key requirements, cost overruns, lower quality, reduced profits, and a shorter productive lifecycle.

5.5   Benefits of Agile Hardware Product Realization

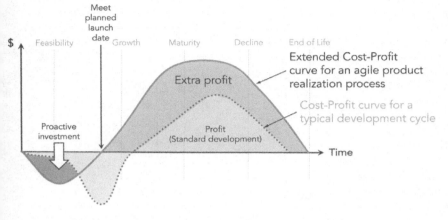

Figure 5.4: Minimizing costs and extending profits with agile hardware.

As shown in Figure 5.4, companies that follow an agile hardware product realization process invest more resources early in the process but realize overall savings over the life of the product. They better understand and utilize resources with a more agile and effective process, and thus can realize the following business benefits:

- Ship to the market on time.
- Ensure the product meets intended requirements.
- Exhibit lower total costs and higher profits for a longer period.
- Achieve higher reliability and more serviceable products.
- Establish robust and resilient supply chains.
- Experience an extended product lifecycle.
- Strengthen their brand image.

6   Preparing for Agile Hardware Implementation

Just as a house needs a solid foundation before you start to frame it, some foundational principles are shared here that should be taken into consideration before you apply the 10 best practices of agile hardware product realization to your own product development journey.

One of the first projects I participated in after starting the PRG was for an innovative professional digital video camera company. The founder was a high-profile billionaire who had purchased a working film studio in Los Angeles. This new innovative camera faced all sorts of challenges – there were a bunch of pre-orders on the books, the committed shipping date was several months late, failures in the factory were high, the development budget was blown out of the water, and the product cost needed to come down significantly to hit desired profitability. Because the product was so innovative and compelling, the backlog of orders kept growing, and the pressure on us to get the product into volume manufacturing kept increasing like a pot ready to boil over. After four months of intensive effort that ranged from Design for Manufacturing (DfM), to improving production processes and yields, to developing automated tests, and to passing regulatory compliance, the product was finally brought to market, and it ended up being a hit for both film studios and prosumers.

For me, the lesson that was reinforced from my Raynet days was that, if the company had a better understanding of NPDI, it could have saved months of time, millions of dollars, and many sleepless nights. There was nothing fundamentally wrong with the product design, but because the business was working ad hoc in development and not engaging downstream team members early as part of a strong NPDI process, millions of dollars were wasted in bringing the project back on track, and, as a result, it was over budget, of lower quality, and arrived late to market.

6.1   10 Critical Warning Signs of a Broken Process

If you have concerns about your own process but are not sure what warning signs to consider, take a look at the following list for some of the most impactful:

1. **Slipping the schedule:** Your actual schedule is not meeting your planned schedule. This may result from unclear product requirements, lack of resources, misallocation of resources, a lean budget, feature creep, and/or changing requirements.

2. **Exceeding the budget:** This results from underestimating all costs associated with product development and regulatory and test processes and is often combined with a slipping schedule.

3. **Exceeding your product cost target:** This is often caused by underestimating costs associated with the product. Aspects such as production test, volume materials pricing, packaging, and labeling are often overlooked.

4. **Missing requirements and/or feature creep:** When the product requirements are not clearly defined, the definition of the product during the development process will "creep," which helps to create the previous warning signs.

5. **Lack of critical resources and/or rapid team growth:** This is common in well-funded startups, where it is not possible to hire resources fast enough to meet plans or hiring resources occurs faster than the ability to onboard and integrate team members effectively.

6. **Breakdown in communications:** Communications breakdowns are often tied with rapid team growth and results from cultural misalignment from new hires and/or **unrealistic expectations that impact communications.**

7. **Customer "beta" test failures:** Such failures are common when development schedules are tight and verification plans are thin. This usually means a product redesign is needed (see impacts above).

8. **Low manufacturing yields:** Low yields are found when ramping volume manufacturing and often results from design-related issues and/or a rushed product realization process.

9. High number of pre- and post-launch engineering change orders: These are indicative of a weak development process that has not fully incorporated product requirements into the design nor comprehensive validation testing.

10. High number of field failures and product returns: This is the last place that you want to see product failures as these will directly erode your client satisfaction and brand quality. Such returns indicate a broken product development and validation process.

If you are experiencing a few of these, you should take mitigating actions. If you are experiencing more than five, then your product is likely to be at serious risk.

6.2 Marketing's Role in the Product Realization Cycle

Marketing plays a role throughout the hardware lifecycle, whether this be in the form of the creation of customer needs for the product through market research and writing the MRD early in the cycle or by providing support to ongoing development activities through to finally launching the product to the market once it is complete. Figure 6.1 shows these activities.

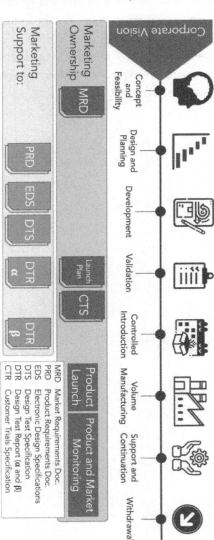

Figure 6.1: Marketing's role in the product realization cycle.

6.2.1 Understanding Your Customer

The first step toward finding product–market fit is to identify and understand the needs of your customer and then to define how the product will meet those customer needs. To launch a successful product, you must first gain clarity about the items shown in Figure 6.2.

1. Who will buy your product?

Ask yourself whether you have fully defined and prioritized your target buyers?

2. Why will they buy your product?

Have you identified the problems they need to solve?
Have you defined the primary customer use-cases that the product will address?

3. How does your product compare to the competition?

Why do you think your product is better or different to competing alternatives?

4. What is your story?

Ask yourself how your product becomes relevant to your target customers

5. How can you reach your target customers?

Make sure you really understand your optimum marketing and communications mix

6. Do you understand the true market window?

Make sure you know when you have to ship a completed product, so it is ready for customers to buy and that you are not too late.

Figure 6.2: Requirements for marketing clarity.

Typically, this information should be included in an MRD, which is a foundational document that establishes the "who" and "why" for the product. The MRD will inform the creation of clear product requirements.

Most technology-based hardware companies put too much emphasis on technology, features, and functionality without fully exploring and understanding market needs. I've heard statements such as "This technology is so cool; it will sell itself." Unfortunately, products don't sell themselves and companies that think they do don't last long. Consider Apple; it has been so wildly successful because it solves real market problems with simple and innovative solutions as well as provides a highly focused portfolio of products. Because of this focus, the company is now setting the market expectations with customers waiting for its new products rather than Apple having to predict market needs. Furthermore, the technology it deploys as part of setting these market expectations is a byproduct that supports its business to best solve real market needs over time.

6.2.2   Defining Your Product Requirements

Once the "who" and "why" market requirements are understood and defined in the MRD, the next step is to define the "what" in the form of a PRD. The PRD defines specific hard requirements of the product that will be designed to fulfill the requirements of the MRD. Whereas the MRD may be somewhat aspirational, the PRD requirements must be measurable and testable. The product is not ready to ship until the PRD requirements have been met.

To illustrate the relationships between the market requirements in the MRD and the design requirements in the PRD, refer to Figure 6.3. This shows a partial listing of requirements in each document and how they relate to one another. The overall product description is as follows:

The operating room black box recorder is intended to record the status of anesthesia equipment, patient monitors, and audio and video procedures.

MRD		PRD		
User Req. Number	Description	Functional Specification Number	Description	Related User Req. Number
URS-01	Equipment must mount on a 19 in rack or on a mobile cart	FS-01	Equipment will be less than 19"x19"x19"	URS-01
URS-02	Equipment must be able to record real time data	FS-02	Equipment will weight less than 20 lb	URS-01
URS-03	Equipment must not be too loud to distract surgeon	FS-03	Equipment will not exceed 30 dB peak volume	URS-03
URS-04	Equipment must be able to record ECG data	FS-04	Equipment will contain storage with at least 1 TB of capacity	URS-02
URS-05	Equipment must be able to record anaesthesia levels	FS-05	Equipment will have a visible screen GUI (4"x4" minimum)	URS-19

Figure 6.3: Example relationships between MRD and PRD requirements.

6.3   Defining the Product

6.3.1   Designing and Documenting

When launching a new hardware product, you need to begin with the end in mind and understand what the product will look like when you bring it to market. The first step in this process is to create a product tree, which is a hierarchical representation of how the product structure is defined. Figure 6.4 shows a representative (high-level) sample of a laptop computer system and the product tree that goes with it.

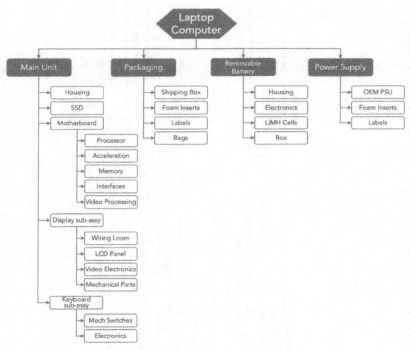

Figure 6.4: Example of a product tree diagram.

To enable your teams to more clearly understand how the product tree relates to how the product is being physically built, this product tree is often augmented with a diagrammatic drawing (exploded view) of how all the pieces fit together, an example of which is shown in Figure 6.5.

Parts List	
A	Display sub-assy
B	Keyboard sub-assy
C	Touchpad
D	Top Case
E	Motherboard
F	Interface Card
G	Battery
H	Hard Disk
I	RAM Module
J	Bottom Case

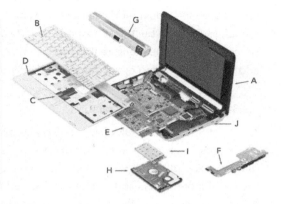

Figure 6.5: Example of a product tree augmentation diagram.

6.3.2   Agile Hardware Product Development Process

The keys to agile hardware development are

- identifying areas with high technical risk;
- identifying areas of high schedule risk, such as material long lead times or long testing times;
- identifying areas with a high market, user, or customer acceptance risk; and
- planning the development schedule to be able to build, test, and purchase materials to mitigate these high-risk areas as early in development as possible.

Without strong technical project management, developers tend to focus on the easy parts first, creating fast progress early in the program, only to hit "unforeseen" technical or supply chain obstacles late in the product. A strong agile technical program manager will prioritize the difficult and risky work first and spend as necessary to mitigate those risks as early in the project as possible.

Modular subsystem design can help isolate risks and allow testing to proceed on individual modules without waiting for availability of other modules. Purpose-built test fixtures and detailed modular interface specifications can help ensure that modular development runs in parallel and that agile learning cycles can be applied independently.

6.3.3 Defining a Bill of Materials

A Bill of Materials (BOM) is a comprehensive list of parts, items, assemblies, subassemblies, intermediate assemblies, documents, drawings, and other materials required to create a product. The BOM is the list of ingredients used to create a finished product, presented in a hierarchical format. Depending on the product, it may include mechanical (hardware) parts, electrical parts (e.g., integrated circuits and Printed Circuit Board Assemblies (PCBAs)), software and/or firmware, and related documents and drawings.

The engineering BOM (eBOM)

- defines all possible configurations (e.g., option classes);
- is, generally, not focused on procurement and manufacturing realities;
- should be created as early as possible in development and kept up to date as the development advances;
- should include cost estimates, targeted suppliers, and expected lead times at volume from production suppliers; and
- should have a revision control process in place so that sample builds and testing are built according to the correct BOM versions.

The manufacturing BOM (mBOM)

- is configuration specific,
- drives the procurement process (manufacturing resource planning),
- supports the manufacturing process, and
- should include manufacturing and test process documentation.

Figure 6.6 shows an example of an overall BOM for the laptop computer example used in this section.

PROJECT NAME	Laptop
PROJECT NUMBER	LT-021-09
DATE	
REV	C

Item	Description	Qty	Purch. Part Number	Material	Supplier	Ea. Cost	Extended cost	NOTES
1	Packaging	1					$224.18	
	Shipping Box	1	XYZ-10002	Cardboard	Box.co	$2.00	$2.00	Custom
	Foam Inserts	2	XYX-10-220	Expanded foam	Box.co	$1.00	$2.00	Custom
	Anti-static bag	1	Baggo-001	Poly	Bag.co	$0.25	$0.25	
2	Main Unit	1				$220.93	$220.93	
	Bottom Case	1	OCD100934	PC/ABS	Case.co	$10.00	$10.00	
	Insert, Brass, M2 x xL	10		Brass		$0.10	$1.00	
	Motherboard Assy	1	PCBA-01232		CM	$143.45	$143.45	Our design
	SSD - 2TB	1	SSD-01-0234100		Mem.co	$30.00	$30.00	
	Display Assy	1	701W-X2/04		CM	$23.43	$23.43	Our design (mech)
	Keyboard Assy	1	Key/01-43.5500		Keys.co	$13.90	$13.90	OEM design
	Screws M2/4mm	10	100.334.6410	Brass	Screw.co	$0.05	$0.50	
2.1	Motherboard Assy	1				$182.75	$182.75	
	Processor Intel Core	1	i7-13700-K		Intel	$61.23	$61.23	
	PCB	1	PCB10/2009.02		PCB.co	$83.56	$83.56	CM Source
	GMCH (northbridge)	1	I82810 GMCH		Intel	$14.96	$14.96	
	Clock Generator	1	Si5351A-B07321-GT		Skyworks	$1.10	$1.10	
	...	..	..		..			
	...	..	..		..			
2.2	Display Sub-Assy	1				$21.90	$21.90	
	Panel, UI	1		PC/ABS		$1.00	$1.00	
	Window, Display	1		PMMA		$0.40	$0.40	
	Display Module	1		Poron TBD		$20.00	$20.00	
	Gasket, Side	2		3M 468MP		$0.50	$1.00	CM quotes
	Adhesive, Window	2						

Figure 6.6: Example of a typical BOM for a laptop computer (partial).

6.3.4   What Should Be Included in Your BOM?

To properly document your product, you should include specific pieces of product data in the BOM record. Whether you are creating your first BOM or are looking for ways to improve how you create one, here is a high-level list of essential information to include:

- **BOM level:** Assign each part or assembly a number to detail into what level it fits in the hierarchy of the BOM. This allows anyone to understand the product structure.

- **Revision number and letter:** Every time the BOM changes (or a new sample is built), the revision level is incremented. During fast agile cycles, it is critical that the configuration of test samples are tracked. Modules and subassemblies should have independent revision levels.

- **Part number:** Assign a part number to each part or assembly to reference and identify parts quickly. It is common for manufacturers to choose either an intelligent (i.e., one that has meaning tied into the part number) or nonintelligent (i.e., just a number) part numbering scheme. You should never duplicate part numbers.

- **Part name:** Record the unique name of each part or assembly. This will help you identify parts more easily.

- **Description:** Provide a detailed description of each part that will help you and others distinguish among similar parts and identify specific parts more easily.

- **Quantity:** Record the quantity of each part number used in each assembly or subassembly to help guide purchasing and manufacturing decisions and activities.

- **Unit of measure:** Classify the measurement in which a part will be used or purchased. It is common to use "each" for discrete items, but standard measures such as inches, feet, ounces (or millimeters, grams, etc.), and drops are also suitable classifications. Be consistent across similar part types.

- **Phase:** Record what stage each part is at in its lifecycle. For parts in production, it is common to use a term such as "In Production" to indicate the stage of the part. New parts that have not yet been approved can be classified as "Unreleased" or "In Design." This is helpful during the NPI because it allows you to easily track progress and create realistic project timelines.

- **Procurement type:** Document how each part is purchased or made (i.e., off-the-shelf or made to specification) to create efficiencies in manufacturing, planning, and procurement activities.

- **Reference designators:** If your product contains PCBAs, you will need to include reference designators that detail where the part fits on the board in your BOM.

- **Estimated BOM cost:** An estimated cost of the total quantity of parts on the BOM. Regularly updating this will help catch cost-creep early.

- **Estimated volume lead time:** Estimate the lead time of parts that are purchased in production volumes to help with planning of test cycles. Having these lead times available might require advance risk buy purchases.

- **Notes:** Capture other relevant notes to keep everyone who interacts with your BOM in the know.

Depending on your product and business needs, you may include additional information (e.g., drawings, specifications, firmware, etc.).

All of the above and some other characteristics create what is essentially an item master record for each part that you use within the product. This can be viewed like the DNA of a piece or part, because all of the elements that make up the item are attached to the item and live with it for the life of the product. Long after the design engineers have finished their work on a product, manufacturing engineers can use this DNA to understand how to build the product in manufacturing and even find replacement parts, as necessary, with the hope of not having to perform a redesign. Figure 6.7 shows a representation of this item master in the form of the part "DNA."

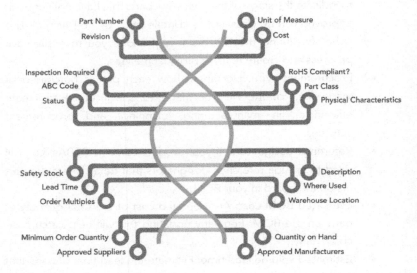

Figure 6.7: Elements of the item master: BOM part "DNA."

6.3.5   Importance of Selecting Parts and Materials Early in Design

Not too long ago, a medical device client we had worked with inadvertently selected a hard-to-get part with a 52-week lead time and incorporated it into its new product design. Regretfully, as a result, a planned production date of February got delayed until November while the company waited for this part to arrive. Unlike software, a single missing hardware part (even a screw) can bring your production to a halt. This nine-month delay could have been avoided if the proper materials risk analysis and planning had been performed earlier in the design process.

To move your product from the lab into full-scale production, you should define all the parts and materials that go into your product and gain inputs from your procurement and supply chain team as early as possible to ensure that parts are easily available, have duplicate sources, are not approaching their end of life, and meet quality and reliability requirements as well as cost targets.

Once parts are locked into the design, it is extremely difficult to change. Moreover, this issue is compounded if you are in a regulated industry, for example, one in which you must go through a formal FDA recertification effort to validate changes to your product. According to Aberdeen Group, only 12% of companies involve sourcing and suppliers before the prototype and pilot phases.[2]

The graph in Figure 6.8 shows how early decisions on suppliers and sourcing can impact the overall cost of the manufactured product.

For a deeper look at materials and procurement practices, Section 7.5 Apply DfX.

---

[2]https://www.prolim.com/wp-content/uploads/2017/11/Procurement-in-New-Product-Development-Ensuring-Profit-From-Innovation.pdf

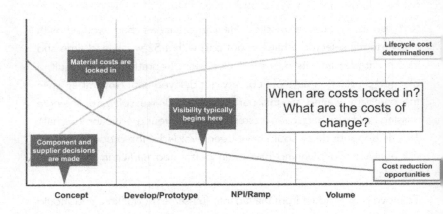

Figure 6.8: Effects over time on the locked-in costs of a product.

6.3.6   Applying an Agile Process to Parts and Material Selection

One of the biggest obstacles to using agile methodology in hardware design is the long lead time of securing components as well as the time needed to assemble and test them as part of the product. Some hard-to-get component lead times can be a year or more, and pilot assembly and testing can easily add weeks to months more.

The implementation of rapid hardware sprints requires upfront planning and a commitment to evaluating, selecting, and qualifying strong vendors early and then purchasing materials at risk (for early testing and to secure long-lead-time materials) that might be needed for future pilot production runs. In waterfall project plans, materials are procured cautiously whereas, in agile hardware planning, materials are purchased early and aggressively with the understanding that some materials purchased "at risk" will not make it into the final design and will need to be scrapped. Materials for the highest risk parts of the design should be purchased, prototyped, and tested as early as possible. Design and procurement decisions for long-lead-time materials should also be front loaded. Consider leveraging larger distributors of electronic components to provide a comprehensive BOM analysis and samples for limited prototype runs. This can save time early in the development process and ensure a smooth supply of materials later in production.

Good materials planning and procurement of high-risk parts can save several months on a product development and launch schedule.

6.4   Tradeoffs among Cost, Performance, and Time

When developing physical products, you will invariably need to make tradeoffs. You need to understand the tradeoffs required to bring your product to market and to make decisions that will be best to support your creation of long-term business value and growth. For example, if you want to accelerate your schedule, you will likely need more resources, which will increase your costs as well as your development and quality risk. If you want to add features, then your resource needs, cost, and time will all increase.

The rule of thumb is that you generally can pick only two, so the third one will shift. In an agile hardware development process, the focus is on fast build and test cycles at the expense of increased cost of materials and, if necessary, a reduction of product performance and feature goals so that a Minimum Viable Product (MVP) can be delivered quickly, as illustrated in Figure 6.9.

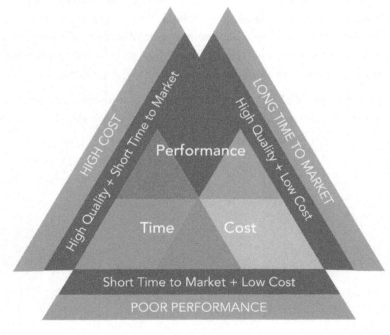

Figure 6.9: Performance triangle.

6.5 Accounting for "Total Cost of Ownership"

Just like a pile of wood on your lawn does not equal a house, a pile of parts on the factory floor does not equal a shippable product. When developing a hardware product, it is important to understand your total product cost in addition to the most visible costs, such as material and manufacturing costs.

6.5.1 What Makes up Total Cost?

Total cost quantifies all costs associated with a product including overhead costs and allocates costs to the appropriate product and is calculated using the simple formula

$$selling\ price - total\ cost = true\ profit$$

Total cost is the only way to compute the real profitability of a product as it provides an objective basis for product portfolio planning (prioritizing products with the highest return), enabling calculation of the expected return from given resources as well as forming the basis for rationalizing the elimination of money-losing products. In the absence of a total cost view, development resources have a greater chance of being wasted on unprofitable products.

Alternatively, by using total cost as your measure, you will promote behavior that minimizes total cost (versus optimizing cost in a particular area such as engineering). In addition, a total cost view will prioritize development resources for the highest return products, while eliminating the case in which profitable products are subsidizing unprofitable ones.

6.5.2   Basic Elements of Total Cost

Figure 6.10 shows all the costs that should be included when scaling a hardware product business.

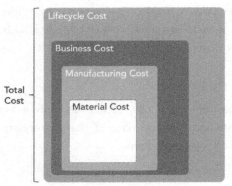

Figure 6.10: Elements of total cost.

It is common for companies to underestimate total cost by limiting their calculations to the cost of goods sold. When calculating a product's total cost, you should include all of the following elements:

- material cost;
- manufacturing cost:
    - labor and
    - overhead (e.g., indirect labor, depreciation, utilities, and facilities charges);
- business cost:
    - marketing, sales, and royalties,
    - finance, human resources, legal, and administrative,
    - engineering,
    - insurance,
    - capital equipment,
    - other costs (e.g., inbound freight, systems, carrying costs, and scrap); and
- lifecycle cost:
    - support and
    - warranty (repair, replace, and recall).

6.6   Impact of Cost Escalation throughout the Product Lifecycle

As you go through the key phases of hardware NPDI, the costs associated with each phase increase. Therefore, identifying an issue in feasibility is much less expensive than discovering the same problem during volume manufacturing. This is considerably different from the impact for software products because software code may be replicated across multiple installations without the need to purchase material, build physical products, and ship them to customers.

For a hardware example, let's look at the Samsung Galaxy 7, which was prone to catching fire in the field and eventually recalled. If Samsung had identified and mitigated this risk during the feasibility phase versus after shipping the product during volume production, it could have eliminated its $5.3 billion dollars in recall costs, brand erosion, and safety risk that resulted from this defective design getting into the market in high volume.

Early customer deployments with real-world feedback and analysis combined with intentional scaling (from 1 to 10 to 50 to ...) will help you to identify and mitigate risk early and minimize liability in volume production. Figure 6.11 highlights the relative scale of this escalation as you move through the lifecycle of the product.

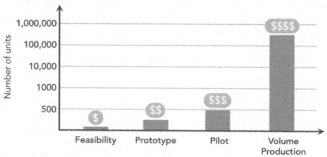

Figure 6.11: Cost escalation through the product lifecycle.

The impact to your business of cost escalation is huge. Therefore, addressing potential issues early is critical for a successful product launch.

6.7 Flexibility Versus Control through the Hardware NPDI Process

In a strong NPDI process, you will want to support maximum flexibility early in the development process because research scientists and engineers need to have the flexibility to experiment during the concept and feasibility phase, where ground-breaking ideas are formed.

However, once the product definition and MVP have been formalized, more process controls need to be added into the mix to ensure that, when the product gets to manufacturing, the documentation will be in place to show production assemblers how to build the product in volume without handholding from engineering. This differs dramatically from how software products following an agile development methodology are controlled because there are no parts to procure or assembly involved. For software products, all it takes is to hit a button and release the latest version of code into the field and all instances will be updated automatically. This is a key difference between how agile software and agile hardware are deployed in practice.

As you gain clarity about the product definition and want to build more than just laboratory prototypes, you will want to implement automated business systems such as Product Lifecycle Management (PLM) systems and Quality Management Systems (QMSs) to enable better control and communication of your product data across geographic and organizational boundaries. Figure 6.12 illustrates how this balance between flexibility and control shifts between the early concept phase and the point of release to full-scale manufacturing.

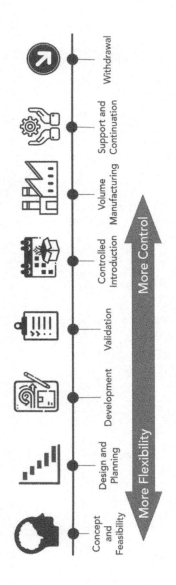

Figure 6.12: Moving from flexibility to control during the hardware lifecycle.

7 The 10 Best Practices of Agile Hardware Product Realization

NPD New Product Development
NPI New Product Introduction

Figure 7.1: The 10 best practices of agile hardware product realization.

Once you have prepared for agile hardware product realization, you are now ready to apply the 10 best practices, as shown in Figure 7.1. The 10 best practices are

1. Employ agile hardware development.
2. Leverage simulation tools and rapid prototyping.
3. Develop an MVP.
4. Understand and mitigate risk early.
5. Apply Design for Excellence (DfX).
6. Incorporate product reliability, validation, and testing.
7. Meet agency and environmental compliance.
8. Deploy scalable business systems and processes.
9. Develop a resilient supply chain.
10. Verify readiness for volume manufacturing.

By following the 10 best practices described in this book, you can realize substantial business benefits and bring better products to market faster, at lower cost, and with less risk.

While these 10 Best Practices are shown in numerical order, and roughly follow the product development flow, they should be considered highly integrated and complementary activities that should be implemented holistically and in a parallel fashion to best support your business goals.

7.1   Employ Agile Hardware Development

Agile hardware development takes the best principles and practices from the Agile software development model and applies them to the creation of physical, hardware-based products. Depending on your existing development environment, you may need to adapt these agile principles to work within your unique product and market constraints as well as deploy them over multiple product development cycles to iteratively improve your NPDI process. Some key elements of agile hardware-based product development include:

- **Cross-Functional NPDI Team:** Effective hardware development necessitates a cross-functional team environment, where different roles and skillsets – including design, manufacturing, testing, service, etc. – collaborate and communicate regularly through the PLC process. This fosters innovation, reduces risk, and enables a faster exchange of ideas and quicker problem-solving.
- **Iterative Development "Sprints":** Agile hardware development follows a rapid, collaborative, and iterative process. This iterative process allows for regular feedback and real-time adjustments throughout the product development cycle.
- **Daily meetings:** The project team has a short meeting every day to discuss progress, upcoming work, and any potential hurdles. It gives the team space to discuss any challenges that could slow down the project, and the ability to brainstorm solutions, or get feedback from other experts on the team.
- **Real-time Collaboration:** Empowered, cross-functional teams are the backbone of agile product development. Enabling software tools (generally SaaS based) provide real-time collaboration across organizational and geographic boundaries. Instead of a design team work on a prototype, finalize it, and then show it to the manufacturing team, you have the two perspectives working in concert with each other, which can uncover challenges early in the process.
- **Customer Collaboration and Feedback:** The objective from this collaboration is to create a product that best serves the customer's needs, so their input is sought at every stage of development. Regular demos and prototypes are shown get

customer feedback and make necessary design adjustments. This is where simulation and rapid prototyping tools are useful.

- **Modular Design:** Given the physical nature of hardware, changes can be expensive and time-consuming. To reduce modifications, a modular design approach should be applied, where different parts or components of the product are designed and built separately. This allows for changes to be made to specific modules without affecting the entire product. In addition, a stable hardware platform should be specified, where field modifications can be achieved via software updates vs. making hardware changes after the product has shipped.

- **Simulation, Rapid Prototyping and Testing:** These powerful tools are key to a more agile framework and enable greater speed, risk identification, user feedback, support DfX as well as fosters collaboration early in the development process.

- **Flexible Planning:** Plans in agile development are more flexible and evolving as new information is gathered from each sprint. Instead of trying to plan everything upfront, plans are revisited and revised regularly based on the progress and changing needs within the framework of higher-level project milestones and deliverables.

- **Value-Driven Delivery:** The goal is to deliver the most valuable features first, prioritizing work based on the value it delivers to the customer. Understanding and mitigating risk early should also be a key consideration when prioritizing and sequencing activities.

- **Retrospective Meetings:** After every sprint, the full team meets to discuss what went well, what didn't, and what can be improved in the next sprint. This constant process of introspection and iterative learning helps the team improve over time.

In conclusion, agile more of a mindset than a rigid set of rules. The idea is to create a culture of collaboration, flexibility, continuous improvement, and customer focus, and the specific practices can be adapted based on the nature of the product and the team's requirements. Also, once design freeze has happened, the shift to a more structured and controlled process will be necessary (for medical devices, once design controls are in-place), otherwise, there is real risk of data corruption between the agile development and the supporting supply chain, which may result in the wrong parts being purchased and/or the wrong product being assembled.

7.1.1   Select Your Agile NPDI Team

Getting the right people involved early will help to support downstream success for your product, so you will want to create a dedicated cross-functional NPDI team at the beginning of the project. This team is committed to the project rather than engaged on a full-time basis. The team includes the following members:

- **An experienced NPDI project manager** drives the project. This individual defines the NPDI schedule, including deliverables, task linkage, due dates, and owners. The project manager also tracks decisions and impacts throughout the NPDI process.
- **Marketing personnel** ensure that there is demand for the product and that it meets market needs.
- **Engineering team members include** industrial design, mechanical, electrical, and software engineers. They ensure the product's technical feasibility. Reliability, quality, and compliance engineers support parts selection and testing that verifies the product will perform to specifications throughout its lifecycle.
- **Operations and supply chain team members are** responsible for quality, systems, cost, and delivery. They ensure that the product can be cost effectively built, tracked, shipped, and supported throughout its lifecycle.
- **Financial experts are** responsible for budgeting, tracking and reporting on projects to achieve financial targets.

- **Service and support personnel are** responsible for installation, training, support, field service, and reverse logistics such as return materials authorization replacement and repair.

7.1.2 Create the NPDI Project Schedule

Always begin with the end in mind. Developing a comprehensive and realistic schedule will give you a solid understanding of what is required to get your product to market on time and within budget. If this initial schedule is created with full team inputs, it is likely that you will identify and mitigate potential roadblocks before they become problematic. In addition, this schedule should be managed as a "living" document, which should be adjusted dynamically as the NPDI process progresses through the various stages of development and commercialization. The NPDI project schedule should include the following:

- **Key functions and activities: These include**
    - development (e.g., industrial design, mechanical design, electrical design, firmware, and application software);
    - regulatory and standards compliance testing (e.g., UL, EMC, FCC, FDA, RoHS, and WEEE);
    - verification and validation testing (along with DfX reviews), involving engineering, design, and production;
    - tooling;
    - packaging and labeling; and
    - supply chain and materials.
- **Dependencies:** Hardware inherently has more dependencies than software, so you will want to incorporate these into your schedule and understand them early.
- **Resource allocation and loading:** Resource time required will vary across project phases, so understanding what resources are needed when will help to meet your schedule and budget.
- **Project costs:** Project costs include all the costs associated with the project including resources, materials, and outside support.
- **Phase-gate reviews:** For each phase of iterative development defined, there are critical decision points that need to be addressed before moving forward with the next iteration of development.

Figure 7.2 shows a representative section of a typical product schedule.

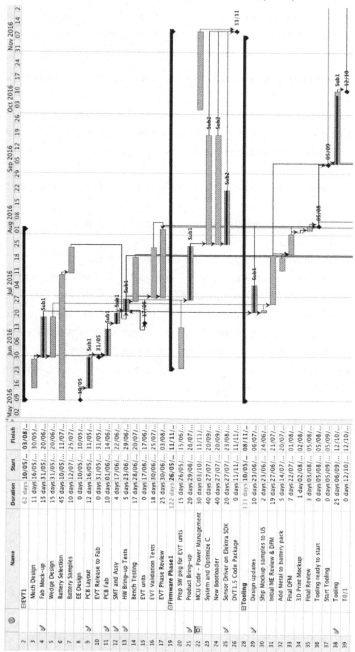

Figure 7.2: Example of NPDI product schedule.

7.1.3   Define Your Product Profile

There are many factors that go into deciding on a product realization strategy. These factors start with the dynamics of your product, market. and the business. Some key factors to consider include the following:

- **Industrial design:** How will the final product look and feel to the customer?
- **Complexity:** How easy is the product to build and test in volume?
- **Cost:** How critical is the hardware cost of goods sold to your business's profitability (versus recurring software or services revenues)?
- **Control:** How much control do you need to have over your product and process?
- **Volume:** How many units are you planning to ship (per month, quarter, or year)?
- **Location:** Where do you intend to ship your products?
- **Quality and reliability:** What level of quality and reliability do you need to meet? In what environmental conditions do your products need to operate without failing? (For example, medical and automotive products generally have more stringent quality and reliability targets than lower cost consumer products.)
- **Expected life:** How long does your product need to operate in the field before being replaced?

7.1.4   Select Your NPDI Strategy

Depending on your unique product profile and business needs, you should select an NPDI strategy before you start the product development process. In addition, product configuration needs to be determined and defined early in the process. This includes identifying the number of products, variants, options, and accessories. Think of this from a customer perspective – that is, how do customers identify the end product that they want and what comes in the box when delivered. This has a direct connection to how you will approach order fulfillment (e.g., "mix and match" versus "off the shelf"). The basic NPDI strategy encompasses the following:

- **Development and validation: These tasks are performed** in-house or in-house mixed with external consultants. External development is performed with an internal project manager.
- **Manufacturing and fulfillment: These tasks are performed** in-house and in-house for final assembly with external subtier suppliers. Outside third-party logistics (3PL) is deployed for final assembly with external subtier suppliers. Full external turnkey manufacturing included direct fulfillment.
- **Combination of development and manufacturing:**
  - Original design and manufacturing: An external firm supports both development and manufacturing. (Note that there are often intellectual property issues that come into play for this model that should be considered before selecting it.)
  - Joint development and manufacturing: This includes a balance between the Original Equipment Manufacturer (OEM) and the original design and manufacturing team. Key elements of design are performed by the OEM and other development along with manufacturing is performed by the joint development and manufacturing team.

Key Insight: Agile hardware new product development and introduction

Successful businesses start with the end in mind. Getting the right team members assembled and developing a comprehensive and realistic schedule, along with a clear product realization strategy, will provide you with a solid foundation to effectively navigate the product realization process and successfully launch your new product in the market.

7.2   Leverage Simulation Tools and Rapid Prototyping

7.2.1   Modeling and Simulation to Accelerate Product Development

There are many tools and techniques that can help you to better understand the behavior of your design before you commit to building a physical prototype. By leveraging engineering modeling and simulation (see Figure 7.3), you can substantially reduce your development time and more thoroughly evaluate and iterate on design concepts of the final product, thus identifying and reducing potential risk before investing in physical prototypes.

Figure 7.3: Computer simulation and modeling of complex parts.

7.2.2   Design Simulation Software

Simulation software allows you to perform subsystem modeling for products that are hard to physically reproduce during the design phase, such as a single element of a complex system (e.g., a telecom network).

To simulate your design, you will first need to create a mathematical model of your subsystems as well as your overall system. The model can be expressed as a block diagram, as a schematic, as a state diagram, or even in Very High-Speed Integrated-Circuit Hardware Description Language (VHDL) and Verilog code for Field Programmable Gate Arrays (FPGA) and Application Specific Integrated Circuit (ASIC) designs. Once created, the simulation will demonstrate the behavior of the model(s) as conditions change and evolve over time or as you inject various outside-world inputs. Data displays and three-dimensional (3D) animation will enable you to see the output of the simulation as it runs in real time.

The use of design simulation software allows you to test multiple options for designing your product before you build physical prototypes. This enables you to experiment with alternative ideas in an iterative, real-time way, like agile software sprints.

7.2.3   Electronic Hardware Emulation and Simulation

Electronic hardware emulation makes it possible to replace a hardware element of your design that is not ready for building to your hardware boards by using a computer simulator. This rapid iterative process supports the agile hardware development methodology.

A good example of this is emulating an ASIC design early in the design cycle by connecting computer simulator inputs and outputs to the places on the PCBA where your future ASIC will be located and then testing the system as if it were in place. While performance may vary slightly from the finished ASIC, this emulation enables you to fully test all the interfaces, software drivers, and data transfers to ensure that they will interact with the board to meet your specifications, without having to fabricate an ASIC, which is expensive and time consuming.

When you are ready to commit the design to fabrication, Printed Circuit Board (PCB) design simulation will ensure that interfaces perform as expected when laid out in the physical world, where items such as trace width and spacing are important. Added to this, the use of design rules for PCB layout, as well as any restrictions placed upon you by a potential contract manufacturer, will ensure that your product can actually be built and that it incorporates all the elements needed for manufacturing in volume (see also Section 7.5: Apply DfX).

Most PCB Computer-Aided Design (CAD) systems include elements of the mechanical design resulting from the electronic design, such as the ability to visualize in three dimension the assembled board with models of each of the components on it. This means that you can validate proper footprints, component spacing, potential interference, and "test fit" multiple boards together to ascertain whether there is physical interference between components on the assembled boards. This is especially useful when you are designing multi-board systems where they all have to fit together in a constricted space (e.g., think of the 3D Tetris involved in creating the functionality from multiple electronic and mechanical assemblies that go into a small mobile phone). Once complete, a combination of outputs from the electronic and mechanical design suites can be used to ensure that your design works before investing the time, resources, and money to build physical prototypes.

7.2.4   Mechanical Design Simulation

With many mechanical CAD packages, a 3D model of the intended design can be used to check the fitting of all the pieces that make up your product. The designed enclosure can be "fitted" with the PCB models from the electronic CAD system alongside other mechanical pieces to form a complete model of your product. Using rapid prototyping techniques (as described below), you can then create a complete "space model" of your product. This space model gives everyone on your team, customers, and investors a chance to explore the idea and provide feedback.

7.2.5   Rapid Prototyping

The most common type of rapid prototyping for custom-designed mechanical parts or other nonelectronic parts is 3D printing, which is an additive manufacturing process. It is highly accurate, material compatible, fast, and cost effective in limited quantities (see Figure 7.4[3] for an example of a 3D-printed prototype). One caution is warranted when using 3D printing: In most cases, there will need to be some level of redesign to support production tooling (hard tooling), given that 3D printing allows the creation of geometric shapes that may not be possible to make when designing hard tooling. Ways that rapid prototyping can be used to support agile hardware NPDI practices include the following:

- **Feasibility prototypes (feasibility testing):** These are used for prototyping new technologies to verify that the design is feasible and to help mitigate potential technical risk.
- **Low-fidelity user prototypes (behavior testing):** These are interactive wireframes that don't look like the finished product. They are created to test the user experience with design interactions and to simulate user experience to identify usability issues early in the design process.
- **High-fidelity user prototypes (usability testing):** These are realistic-looking working models that are useful for sharing a proposed product to stakeholders. They are often used in user testing to learn what works and what does not work well.

---

[3]Image: https://www.aha.org/aha-center-health-innovation-market-scan/2022-06-07-3-ways-3d-printing-revolutionizing-health-care

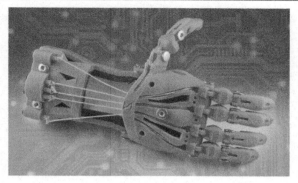

Figure 7.4: Example of 3D printed prototype parts.

7.2.6 Benefits of Leveraging Modeling, Simulation, Emulation, and Rapid Prototyping

Many of the simulation and rapid prototyping tools described above did not exist until relatively recently. These tools now enable agile methodologies to be applied more effectively early in the development process for hardware-based products and enable faster innovation cycles. Benefits of leveraging virtual and physical simulations and emulations of hardware designs include

- substantially reducing product development time and cost;
- improving end-user (and stakeholder) involvement;
- receiving early market validation and feedback;
- exploring a range of designs quickly;
- testing end-part materials early;
- evaluating and identifying potential ergonomic and product hazards;
- reducing the risk of technology, system performance, and fit issues;
- improving product reliability and quality; and
- lowering the rate of production and field failures.

Key Insight: Leverage simulation tools and rapid prototyping.

The more you can leverage simulation and rapid prototyping tools and techniques, the faster you can move through iterative development cycles. In addition, following a layered approach for development and testing will help (e.g., simulate and prototype subsystems individually before bringing them together into the full system).

7.3   Develop an MVP

7.3.1   Importance of an MVP

It is easy for innovators to get excited about a new technology and to think that superior technology and more features will sell in higher volumes than inferior technology with fewer features. However, when bringing new hardware products to market, it is imperative to understand that every feature adds time to your development schedule, as well as complexity and cost to your product. Figure 7.5 gives a graphical interpretation of the definition of MVP.

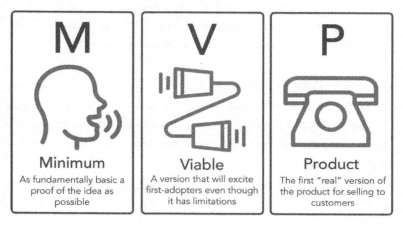

M	V	P
Minimum	**Viable**	**Product**
As fundamentally basic a proof of the idea as possible	A version that will excite first-adopters even though it has limitations	The first "real" version of the product for selling to customers

Figure 7.5: Definition of MVP.

In many cases, the market may not want or even care about these extra features. Remember Google Glass, a futuristic device that leveraged natural language voice commands to power a computer embedded in a pair of eyeglasses. In this case, Google's technology was undeniably "cool" and groundbreaking. However, in reality, customers did not want to pay a premium price for a product that ultimately invaded their privacy. After several negative high-profile press stories that eroded Google's brand, combined with moribund sales, the product was removed from the market and Google's substantial investment in the product was written off.

American entrepreneur Eric Reis wrote in his seminal book *The Lean Startup*, that the MVP is "not necessarily the smallest product imaginable, it is simply the fastest way to get through the Build–Measure–Learn feedback loop with the minimum amount of effort." Therefore, when thinking about an MVP, consider whether you are focused on the most basic requirements:

- Is a feature absolutely necessary for the first product to ship? After all, every additional feature adds complexity to the design.
- Do the product features address the market need without extra "bells and whistles"?
- What are the additional features that are important, but not urgent, that can be addressed after the initial market need has been validated using a strategic product roadmap?

7.3.2   MVP Process

An MVP process is one that you repeat over and over again: Identify your riskiest assumption, find the smallest possible experiment to test that assumption, and use the results of the experiment to course correct, as shown in Figure 7.6.

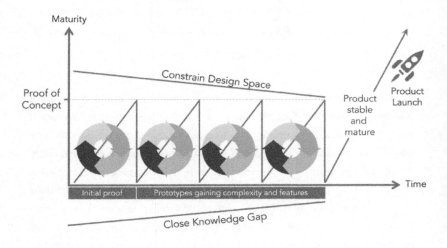

Figure 7.6: The MVP process

7.3.3   What Does a Successful MVP Look Like?

Figure 7.7: Successful hardware MVP: iPhone.

For a successful hardware-based product MVP, let's look at the original iPhone, introduced by Apple in 2007. It is an excellent example of an MVP for the following reasons:

- **Simplicity:** The original iPhone began with a minimum set of features such as making calls, sending text messages, browsing the internet, and listening to music. While Apple could have undoubtably added more features to this initial product, the brilliance of Steve Jobs and Apple was to introduce this minimum feature set to validate the market. Once the market was validated, they progressively add more features over multiple years to grow market share, stay ahead of the competition and reap billions of dollars of extra profits.

- **Customer Validation**: The first iPhone was able to quickly validate customer interest in a new type of mobile device. The massive interest and sales that followed this MVP release demonstrated a clear market demand for intuitive touch interfaces smartphones with internet connectivity.

- **Iterative Improvement**: The original iPhone architecture laid the foundation for continuous improvement and the release of new models with increased capabilities. Based on the feedback from the first version of the iPhone, Apple was able to iterate and make significant improvements to subsequent models. This iterative process, starting from a strong MVP, has helped Apple maintain its leadership position and profitability in the smartphone market.

- **Scalability:** The original iPhone was designed with scalability in mind. They designed a stable hardware backbone and introduced a core software platform that allowed for future improvements and additions, such as app development. When Apple opened its App Store in 2008, the iPhone's functionality expanded tremendously, and its appeal to consumers increased rapidly.
- **Brand Experience:** Even as an MVP, the iPhone encapsulated Apple's attention to design, user experience, and a seamless integration between hardware and software, reinforcing the Apple brand image. This alignment with brand values is a critical component in the success of any MVP.

So, while the original iPhone was much simpler than the versions we see today, it was an excellent MVP example because it presented a product with a minimum feature set that attracted early adopters, validated a market demand, provided a solid platform for future product iterations, was highly profitable from the start, and encapsulated the brand's promise and potential.

> **Key Insight:** Develop an MVP
>
> When thinking about your MVP, in addition to creating a product with the most basic features set, you should think about developing modular hardware subassemblies that can be the foundation for future product iterations, where as much functionality as possible can be done via software. This will save you time and money as you scale your business and expand your product portfolio.

7.4   Understand and Mitigate Risks Early

High-tech hardware-based product companies face myriad risks, all of which need to be considered before investing in a new product launch effort. It is common to focus heavily on mitigating technology and intellectual property risk without fully understanding or considering all the other risks involved. However, companies ignore these risks at their own peril as they will likely return to hurt the business downstream. Consider the following list of risks that you should understand, assess, and develop mitigation plans in the concept phase, before moving into the development phase:

- **Funding:** Have you clearly defined and validated the business model? Do you have a clear understanding of what funding is needed and how you will raise it? Many early-stage companies (and some large ones) vastly underestimate the investment required.

- **People:** Do you have the right people on the project? Have you optimized the use of internal resources (core to your business) versus external (not core to your business) through all phases of the product realization process?

- **Market:** Have you done the upfront market research and discovery to support the creation of a clear MRD and a clear PRD? Is the market opportunity big enough and well enough protected to attract professional investors?

- **Technology and intellectual property:** Have you verified that your technology is viable to support market needs? Do you have intellectual property in place to protect your "secret sauce" from competitors?

- **Development:** Do you have an agile and/or strong product development process? Are you following best practices to support on-time development, achieve cost targets, and meet specifications?

- **Regulatory and safety:** Do you have a clear understanding of the regulatory and safety requirements that your product needs to meet? Do you have technical expertise to perform sufficient pretesting? Have you allocated enough resources and time to support this function through compliance and beyond?

- **Supply chain and manufacturing:** Do you have a supply chain strategy in place prior to engineering development? Is there a plan to mitigate risk for long lead time, single-source, sole-source, and end-of-life and obsolete parts? A single hard to get part can delay your project by many months, so are you prepared to "risk buy" these parts before your design is complete? Is enough time allocated to support DfX through the development process? Have you considered geopolitical issues and component obsolescence in your strategy?
- **Sustainability:** Have you considered product materials, disposal, and reuse as part of your product lifecycle strategy? How can you minimize your carbon and waste footprint as you scale into volume?

Let's pick three examples from the above that highlight the impact of not fully considering and mitigating all the risks. Any of these examples could have fallen short on more than one risk factor, but a single risk factor is being highlighted to make the point.

7.4.1 Mitigating Risk: Funding Example

Skully Helmets is an interesting case study because the company was wildly successful at initially raising money on a crowdfunding platform as well as attracting professional investors early on until it became clear that the claims it made for the product were not technically feasible and that the management team was blatantly mismanaging funds. Technical risk, development risk, and people risk combined to result in missed schedules, not meeting specifications, increased costs, and failing regulatory requirements.

In this case, there were multiple points of failure that caused the funding to dry up. It is common to see multiple points of failure when a startup company goes under.[4,5]

Here is a summary of Skully Helmet's story:

- **Background:** Skully Helmet raised roughly $15 million to develop an Augmented Reality motorcycle helmet.
- **Funding Risk:** The company was not able to raise enough capital to meet the technological claims made during its crowdfunding campaign.
- **Result:** The business went bankrupt. Over 3000 people who pre-ordered never got a helmet and lost their investment.

---

[4]https://venturebeat.com/business/lawsuit-against-ar-helmet-maker-skully-over-strippers-and-sports-cars-has-been-dropped/
[5] https://www.zdnet.com/article/skully-smart-helmet-firm-founders-smacked-with-fraud-lawsuit/

7.4.2   Mitigating Risk: Regulatory Example

Founded in 2006, 23andMe, an innovative genetics testing company, charged ahead with promises to disrupt healthcare as the "world's trusted source of personal genetic information." Unfortunately, that revolutionary language did not pass muster with the FDA and the company was ordered to stop marketing health interpretations for its genetic testing service. This was followed by a class-action lawsuit from people who alleged that 23andMe misled them by promoting health promises without regulatory approval or validated science to back them up. These points are summarized as follows:

- Background: 23andMe pioneered direct access to genetic information for genetic health risk reports. Promising to be the "world's trusted source of personal genetic information."

- Regulatory Risk: A year after submitting applications for seven health reports in 2012, it stopped communicating with the FDA for six months, according to the agency, and prompting it to clamp down. In November of 2013 the company received a cease-and-desist letter forbidding it from selling its spit-in-a-tube DNA test to consumers.

- Result: The company was banned for 2 years from selling its medical-related products. The FDA finally cleared a test for Bloom syndrome after a formal 501K submission, and 23andMe started marketing its test products in the USA again in October of 2015.

With better planning and understanding of the regulatory landscape required for its product, this two-year gap in sales along with the class-action lawsuit could have been avoided.[6,7,8]

---

[6] https://hbr.org/2020/09/23andmes-ceo-on-the-struggle-to-get-over-regulatory-hurdles
[7] https://www.ncbi.nlm.nih.gov/pmc/articles/PMC4330248/
[8] https://en.wikipedia.org/wiki/23andMe

7.4.3   Mitigating Risk: Supply Chain Example

For its 787 Dreamliner aircraft, Boeing increased its risk by trying to save costs and set up a new untested distributed global supply chain to support. Unfortunately, offshoring increased mission-critical risk for Boeing, which resulted in myriad process and quality issues and caused a delay of more than three years in product shipments as well as estimated $2 billion dollars of lost profit.

Clearly Boeing dramatically underestimated its supply chain risk and paid a much higher price for this in missed deliveries, lower product quality, and unhappy customers. [9,10,11]

Here is its story:

- **Background:** Boeing expanded its global supply chain for the 787 Dreamliner without proper verification and vetting.
- **Supply Chain Risk:** Supply chain problems and quality issues caused delays of more than three years.
- **Result:** Estimated lost profits exceeded $2 billion and the company experienced brand erosion in the market.

> **Key Insight:**   Understand and mitigate risks early.
>
> Have your NPI team review each risk as part of a brainstorming session to identify all the potential risks in each area as well as possible strategies to mitigate them. It may take some time, but it will be well worth the effort!

---

[9] https://www.aerospace-technology.com/features/feature1690/
[10]https://supplychaindigital.com/digital-supply-chain/boeing-787-dreamliner-tale-terrible-supply-chain-management
[11]https://www.forbes.com/sites/stevedenning/2013/01/17/the-boeing-debacle-seven-lessons-every-ceo-must-learn/?sh=7f22fd6a15c1

7.5 Apply DfX

Under the label Design for Excellence (DfX), a wide set of specific design guidelines are summarized; examples of these are shown in Figure 7.8.

Each design guideline addresses a given issue that is caused by, or affects the traits of, a product.

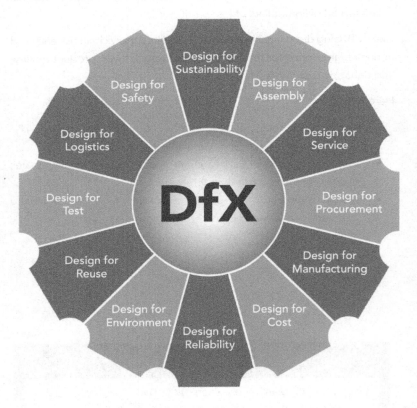

Figure 7.8: Examples of DfX categories.

Common DfX activities include Design for Manufacturing (DfM), Design for Assembly (DfA), Design for Cost (DfC), Design for Reliability (DfR), Design for Procurement (DfP), Design for Test (DfT), Design for Logistics (DfL), and Design for Service (DfS).

It is typical for early-stage companies to struggle with the execution of DfX because they lack enough budget and resources to support the many facets of DfX. If this is the case for your company, then it may be necessary to make a calculated business risk and focus on the DfX categories that are the most essential for the product and business.

Teams that make tradeoffs in its DfX need to document the DfX areas that are left out and note the potential risks that may arise as a result (e.g., more field failures). Other methods need to be applied to de-risk the product and ensure it meets requirements and doesn't put customers or your business at risk.

7.5.1   Allocate Time for DfX in Your NPDI Schedule

DfX will take both resources and time for a hardware product. Enough time in your NPDI schedule must be allocated to support DfX at key product development transitions, as shown in Figure 7.9, to avoid slipping behind schedule.

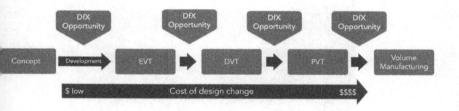

Figure 7.9: Applying DfX through the NPDI process.

The earlier in the development process you can identify and mitigate risk, the less cost impact there will be on the business. For hardware products, especially, the cost of finding and fixing problems downstream can be huge.

7.5.2 Commonly Implemented DfX Types

The following sections expand on some of the most important DfX areas for agile hardware product realization.

7.5.2.1 DfM

DfM is the process of designing parts, components, or products for ease of manufacturing with an end goal of making a better product at a lower cost. This is done by simplifying, optimizing, and refining the product design. Figure 7.10 shows the relationship between the DfM impact and cost of changes to a product throughout the lifecycle, leading to the conclusion that the earlier in the cycle DfM is used, the lower the chance of costly changes being required.

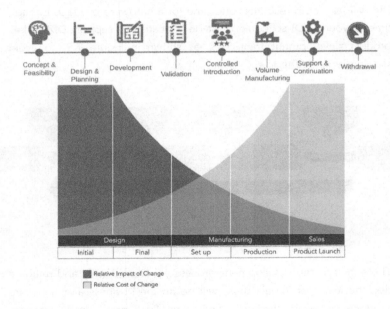

Figure 7.10: Impact of DfM throughout the product lifecycle.

The concept of reducing the number of parts is key to DfM because there is substantial cost associated with each additional part designed into a product. The total cost for each part will include the setup and maintenance of part number and data, planning and procurement, part cost, inbound freight, inventory space and management, inventory carrying costs, part selection, kitting and delivery to production, labor, rework, and scrap, as well as potential excess and obsolescence.

Figure 7.11[12] illustrates the impact DfM can have on the simplification of a mechanical assembly. There are many potential benefits and considerations for each part of the assembly when DfM is applied. Here are some key aspects:

- Eliminate or minimize parts (and part numbers).
- Reuse parts where possible.
- Identify secondary sources for parts when possible.
- Find easily available parts that will not become obsolete within the product lifecycle.
- Build modular designs with logical subassemblies.
  - Allow manufacture of subassemblies with minimal handling and reduced labor.
  - Enable outsourcing.
- Design a core hardware platform and leverage software for as much functionality and as many updates as possible.
- Apply mistake-proofing methodologies, such as poka-yoke (as defined in the following paragraphs).

---

[12] https://logisticsmgepsupv.wordpress.com/2016/04/20/design-for-manufacture-and-assembly/

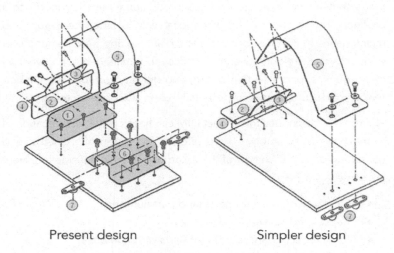

Present design   Simpler design

Figure 7.11: DfM applied to the design of a mechanical part.

Poka-yoke is a Japanese term that means "mistake-proofing." A poka-yoke is any mechanism in a design or process that helps an operator avoid mistakes by preventing, correcting, or drawing attention to human errors as they occur with the intent to eliminate defects.

There are three poka-yoke methods for detecting and preventing errors in production. First, the contact method identifies product defects by testing the product's physical attributes (shape, size, color, etc.). Second, the fixed-value (or constant-number) method alerts the operator to whether a certain number of movements are not made. Finally, the motion-step (or sequence) method determines whether the process steps were followed.

Here are the fundamental poka-yoke guidelines:

- Simpler is better.
- Assembly does not have multiple options – there is only one way to fit the parts together.
- Gauge is not mistake-proof.
- No decision making is required.

Figure 7.12 offers examples of where poka-yoke has been employed to mistake-proof common connector types.

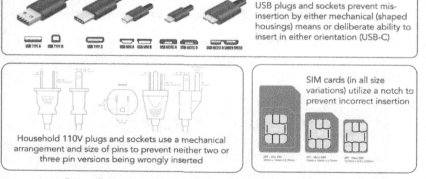

USB plugs and sockets prevent mis-insertion by either mechanical (shaped housings) means or deliberate ability to insert in either orientation (USB-C)

Household 110V plugs and sockets use a mechanical arrangement and size of pins to prevent neither two or three pin versions being wrongly inserted

SIM cards (in all size variations) utilize a notch to prevent incorrect insertion

Figure 7.12: Practical examples of the application of poka-yoke.

7.5.2.2   DfT

Testing is critical to your product's quality and reliability and is easy to incorporate early in the development process with the proper investment and appropriate resources. Companies struggle when they do not plan for production tests early, later finding out that it is impossible to incorporate comprehensive tests, such as in-circuit testing, after the design is complete and the product is in production. DfT comprises a set of techniques employed during the product design phase that allow access to key components during their manufacture as well as adding abilities for self-testing, partial powering of subassemblies, etc.

When I worked at my second startup, SemiPower Systems (an innovative variable-speed motor control company), for example, the engineering development team was in a time crunch and decided to cut two days out of its schedule by not investing the time and resources to incorporate in-circuit test points. This two-day savings in engineering development resulted in a 15-minute production test versus a 30-second in-circuit test.

Enter the issue of scaling. At a volume of 1,000 boards per month over the next three years, the production test resulted in 9,000 extra hours during the manufacturing and test process, not including the extra debugging time. Cutting two days out of the engineering schedule to save a few thousand dollars in engineering had cost the business over $1.5 million and eroded production efficiency and profitability.

Figure 7.13[13] shows a PCBA sitting on a "bed-of-nails" tester, allowing manufacturing test access to many of the components without having to test a complete product.

Figure 7.13: DfT applied to a complex PCBA.

---

[13] Image: (https://www.nutsvolts.com/magazine/article/making_waves)

Here are some tips for implementing a DfT program:

1. Create a comprehensive test plan early in development.
2. Build in test capabilities, where possible.
3. Design in self-test and self-healing, where possible.
4. Develop commands to allow testing during production.
5. Build in diagnostic capabilities for testing and troubleshooting.
6. Design in test points, connectors, and fixture attachment points to support automated production testing.

7.5.2.3   DfP

DfP is a methodology that supports simpler materials procurement, stronger vendor relationships, and lower costs throughout the development and production processes. Redesigning parts after a product is introduced is time consuming and expensive. It is therefore imperative that you select the right parts early in the design process.

Before the design is solidified, there are many ways in which DfP can assist in creating an easier-to-manufacture product. Some key examples include the following:

- Standardize parts in design.
  - Design in standard off-the-shelf parts instead of custom parts.
  - Select easily available parts with multiple sources.
- Minimize the number of parts and suppliers.
  - Select the fewest parts and part numbers possible (see Section 7.5.2.1 DfM).
  - Select the fewest suppliers as possible. Multiple components from the same supplier can reduce cost, improve availability, and simplify management.
- Establish secondary sources of components to the extent possible.
  - This ensures availability (consider leveraging industry software tools for selection).
  - This will provide leverage for negotiating the lowest cost.
- Select strategic suppliers carefully and manage closely.
  - Choose single and sole-source part suppliers.
  - Assign custom-designed parts and subassemblies.
  - Ensure critical technology for your business.
  - Develop vendor evaluation and selection criteria.

-- continued --

- Actively manage supplier risk.
    - Select parts that are not nearing their end of life.
    - Choose suppliers that are financially stable and able to weather a downturn.
    - Allocate sufficient resources to manage strategic suppliers tightly.
    - Lock in costs to prevent increases.
    - Plan availability to support the entire product lifecycle.
    - Develop contracts that support long-term collaborative success.

7.5.2.4   DfL

DfL involves product and design approaches that help to control logistics costs and increase customer service level. These concepts include economic packaging and transportation, concurrent and parallel processing, and standardization. These requirements should be included in both the MRD and PRD as they cannot be revised after the design is complete.

DfL can be used to ensure that your product can be made available in the places and countries where you want to sell it. Some of the options DfL can provide include the following:

- Achieving the lowest weight and size possible will reduce cost and increase transportation options.
- You can design your product to optimize the standard pallet size (which differs for international and domestic shipping). Pallets should be easy to move at reasonable cost and should be constructed of fumigated wood or plastic.
- You can optimize packaging through testing to prevent damage to the product during delivery.

7.5.2.5   DfS

DfS addresses a product's serviceability in the field. Attributes, such as reliability, configuration, and ergonomics, have a direct bearing on the cost and efficacy of servicing the product. Service representatives should be included early in the design process to identify and mitigate any potential serviceability issues in the field. For example, think of a spare tire for a car. This tire enables continued driving until the tire can be successfully replaced. If the spare tire, or temporary repair kit, were not included in the vehicle, then the driver would not be able to drive the vehicle to a service station without towing it.

Product DfS offers a structured framework to help manufacturers determine the right balance between reliability and serviceability. DfS considers customer expectations and competitive products, with the goal of delivering the highest desired level of service performance under given resource and cost constraints.

7.5.2.6 Tips for Implementing DfS

There are several useful DfS techniques that should be considered when you are implementing a program into your process:

- Apply strong DfM and DfT to help with DfS.
- Incorporate design practices that enable customers to diagnose, service, and/or repair products. These should be fully documented; enable self-healing, self-testing, and diagnostics; employ remotely upgradeable software and firmware; and utilize modules that are easy to upgrade.
- Deploy systems that enable remote diagnosis or repair.
- Develop protocols that make field repair faster and easier, with minimal training. Field-replaceable modules should be easy to swap out.
- Ensure that service staff members are trained and knowledgeable prior to product release.

7.6   Incorporate Product Reliability, Validation, and Testing

This best practice has numerous implications for the future quality perceptions of your product, so it is described in two parts: product reliability and product validation and testing.

7.6.1   Product Reliability

Everyone wants the products they buy to work as expected for as long as they use them. However, there is an inherent tradeoff among time, cost, and product reliability that you will need to balance to meet your launch goals as well as product cost targets. Typically, the greater the investment in product reliability, the greater the expense in terms of the cost and time. The graph in Figure 7.14 illustrates these tradeoffs.

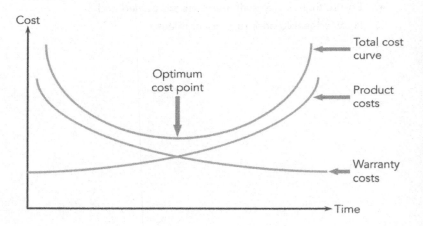

Figure 7.14: Tradeoffs between cost and time for a product.

7.6.1.1 What Is Reliability?

Quality assurance professionals in the American Society for Quality define reliability as the probability that a product, system, or service will perform its intended function adequately for a specified period of time or will operate in a defined environment without failure.

In other words, the primary measurement of reliability is made by the end customer. Therefore, development teams must understand end-user requirements to develop a product that meets customer expectations.

Reliability considerations should include the following:

- probability of success,
- durability,
- dependability,
- quality over time, and
- availability to perform a function.

7.6.1.2 What Are the Risks of a Weak Reliability Strategy?

A weak reliability strategy through the NPDI process will allow potentially defective products to get into the market. Products that negatively impact consumers will also erode your brand. In the worst cases, such as those of automobiles and medical devices, defective products can even lead to injury or death.

Figure 7.15[14] shows the scene on September 13, 2021, of a 2019 Chevrolet Bolt on fire at a home in Georgia, just north of Atlanta. The fire report[15] read "The fire appeared to be coming from a 2019 Chevy Bolt electric vehicle The vehicle was pulled from the garage; however, it had already received extensive damage."

---

[14] https://www.freep.com/story/money/cars/general-motors/2021/10/12/gm-lg-chevy-bolt-recall-battery/6101025001/

[15] Cherokee Country Fire and Emergency Services Fire Report.

Figure 7.15: Example of what can go wrong with reliability.

In this case, General Motors failed to predict that the battery pack, provided by a subcontractor, LG Electronics Inc., could overheat to the point of starting a fire or explosion. The problem stemmed from two small components. First, an anode tab was torn. This is the piece of the negative electrode that wires the cell into a group of cells, called a module, and then into the full battery pack. Second, there was a folded separator – the thin sheet of material that separates the anode and cathode.

The subsequent recall cost General Motors $1.9 billion and involved 141,000 vehicles. The incident also delayed shipment and damaged the brand.

7.6.1.3 Developing a Holistic Reliability Strategy

Reliability principles should be embedded into your company culture and throughout the NPDI process. A holistic reliability strategy will identify potential product reliability risk early in development and reduce the risk of catastrophic product failures in the field.

7.6.1.4 Defining Reliability Goals

Based upon your product profile and market needs, complete reliability goals that include function, characterization of the product's operational environment, probability of success (reliability), and duration should be established. Specific goals needs to be set for key operating elements such as setup or installation and operation in the field. Next, your failure rate goal must be translated into a probability of survival (e.g., 97% of products survive for 3 months without failure). This probability of survival may change over time depending on changing market conditions and customer expectations.

Good reliability goals include

- probability of product performance,
- intended function,
- specified lifespan,
- specified operating conditions, and
- customer expectations.

Once you have defined your reliability goals, you need to understand what the risks and barriers are to achieving the goal. Only then can you incorporate the reliability strategy into your NPDI framework that will enable you to achieve your goals. Some elements you should consider include the following:

- **Reliability model:** This would include a block diagram with a breakdown of the reliability goals for each of the major elements in the system.

- **Reliability plan:** The reliability plan serves to estimate the reliability of each element of the model and compares and adjusts the goals to achieve the desired reliability target. It should also identify areas that do not meet budget targets and so that you can evaluate options that will achieve the product cost target. A good plan helps the team identify the areas of the design of the product likely to fail, so that they can iteratively make design improvements that will support the desired product reliability goals and cost target prior to market launch.

- **Plan implementation:** This enables you to track reliability issues such as defects, bugs, or prototype failures that could cause a system failure and impact customers. When implementing the plan, you need to get the right information to the right people so that they can make better-informed decisions with the best available reliability information at the time.

- **Measuring failures to understand reliability effectiveness:** Documenting problems and tracking progress to the plan comprise the essence of a good reliability program. Some ways you can measure effectiveness include:
  - field failure rate,
  - warranty,
  - actual field return rate, and
  - dead on arrival rate.

7.6.1.5   Leveraging DfR Techniques

There are several reliability tools and techniques that should be identified and implemented selectively, based upon your needs, to help you achieve your product reliability goals as well as to improve your product reliability over time. Some common DfR techniques include the following:

- **Reliability modeling:** The core function of a reliability model is to evaluate an electro-mechanical system to predict its failure rate. The methods used to assess failure rate are described in reliability prediction standards.

- **Thermal analysis:** A precision-controlled temperature program is applied to allow quantification of a change in a material's properties with change in temperature.

- **Derating analysis:** Product reliability testing is performed to help determine whether there are any underspecified components in a product.

- **Tolerance analysis:** Tolerance analysis is used to determine the overall cumulative variation and effect of variation on products stemming from imperfections in manufactured parts.

- **Component and subsystem stress testing:** Deliberate intense stress testing is conducted to determine the reliability of a given component or system. It involves testing beyond normal operational capacity, often to a breaking point, to observe the results.

- **Failure modes and effects analysis (FMEA):** "Failure modes" refer to the ways, or modes, in which something might fail. "Effects analysis" refers to studying the consequences of those failures. Failure modes and effects analysis is a structured way to identify and address potential problems or failures and their resulting effects on the system or process before an adverse event occurs.

- **Highly accelerated life testing:** In this specialized reliability testing, the product is exposed to extreme environmental conditions (stress, strain, temperatures, voltage, vibration rate, pressure, etc.) until it ultimately fails. This information is then fed back into design to improve the reliability of the product during the development phase.

- **Ongoing reliability testing:** Ongoing reliability performance evaluation of your product using samples drawn from production ensures identification of anomalies or changes that may occur in the design, supply chain, or production process that could significantly affect field reliability performance.
- **Highly accelerated stress screening:** Such screening entails stressing a unit to a predetermined level. It can be used as a screening tool to identify high- or low-quality components during the production phase to identify manufacturing process defects that could cause product failures in the field.
- **Root cause analysis:** This process involves discovering the root causes of problems to identify appropriate solutions. In root cause analysis, one assumes that it is much more effective to systematically prevent and solve underlying issues rather than just treating ad hoc symptoms and responding without addressing the true cause of the problem.

Key Insight: Incorporate product reliability.

Reliability is a critical area that should not be taken lightly by your organization – so do not skimp here. Depending on your product complexity and reliability resource constraints, this may be an area where it is more cost-effective to leverage outside experts rather than hire full-time internal resources, because the need is generally transitory.

7.6.2   Product Validation and Testing

For most technology-based products, production tests are the largest investment and biggest bottleneck to scaling your business. An example in-circuit production test station is shown in Figure 7.16.[16]

Figure 7.16: Example of a production test station.

7.6.2.1   Why Are Product Validation and Testing Important?

Product validation is important because it gives you a measurable way to verify that your product meets intended specifications in the development process (via an Engineering Validation Test (EVT) and a Design Validation Test (DVT) and through to a Production Validation Test (PVT)). Product testing is important because it will aid in creating engineering prototypes, result in faster production throughput, raise product yields, make it easier to identify and fix production failures, and reduce field failures and product returns. A combined effective validation and test strategy will lower your costs throughout the product lifecycle.

---

[16]https://acculogic.com/services/test-engineering-services/in-circuit-test-programming-services/agilent-hp-3070-test-program-development-services/

7.6.2.2   What Types of Validation Are Used in the Product
Realization Process?

Effective validation testing begins with a clear understanding of how the product is intended to perform (specified in the MRD, PRD, and Engineering Design Specification (EDS) documents). Validation testing should be designed to verify and validate that the product meets the specifications. For example, if an LED needs to be bright enough to see in the sun, then a good validation test will simulate the light intensity of the sun and "verify" that the LED is visible per the written specification. This will be done for all critical parameters of the product, including software. Consequently, this type of validation needs to have sufficient time and resources committed to support it.

Here are the most common types of validation tests:[17]

- **EVT:** An EVT is performed on initial engineering prototypes to ensure that the basic unit performs to design goals and specifications. Verification ensures that designs meet requirements and specifications while validation ensures that the created entity meets the user needs and objectives. The EVT build is the first time you combine "looks like" and "works like" into one form factor with intended production materials and manufacturing processes.

- **DVT:** A DVT enables you to prove ("validate") that the device you've built works for the end user as intended. FDA 21 CFR 820.3 states that design validation is "establishing by objective evidence that device specifications conform with user needs and intended use(s)." The DVT build should be a configuration of your production-worthy design, made of components from production processes (and hard tools) and on a line following production procedures.

---

[17] Definitions for EVT, DVT, and PVT are inspired by definitions found in Wikipedia.

- **PVT:** The PVT is the final validation phase before a product goes into mass production. Hard tooling is fixed, meaning that no more changes to either the product design or production molds can be made. Jigs, fixtures, and test benches must be in place and validated for the production pilot to begin.

7.6.2.3   Traditional V-Model of Hardware Product Realization

The traditional V-model, which you may have seen or heard of, demonstrates the relationships between each phase of the development lifecycle and its associated testing phase. The phases are shown in Figure 7.17 between the diamonds representing the "phase gates," which provide for risk and cost reduction measures in an NPDI process. The V-model is a representation of system development that highlights verification and validation steps in the system development process and is considered a more constrained way of development than the agile methodology discussed above. This type of development may be more commonly used for highly regulated products (e.g., medical devices and automotive parts).

The left side of the "V" identifies steps that lead to product development, including system specification and detailed system and subsystems product design requirements. The horizontal and vertical axes represent time or project completeness (left to right) and level of abstraction (coarsest-grain abstraction uppermost to most detailed at the lowest level), respectively.

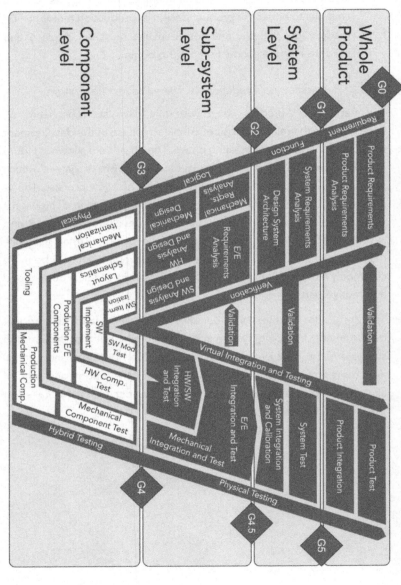

Figure 7.17: Traditional V-model.

Let's step through the V-model, starting with the top (whole product) and ending at the lowest level (component).

Whole product: This is the highest level of abstraction. It includes the creation of the MRD and the PRD along with the product integration and validation test documents to ensure that the product meets the specified requirements.

System level: This is the next level of abstraction. It includes the system-level analysis as well as the supporting system-level architecture and requirements on the right side. The right side includes creation of system-level integration, calibration, and testing plans to ensure that the system meets the desired requirements.

Subsystem level: This layer includes on the left the detailed analysis and development documentation to support engineering development activities (including electrical, mechanical, and software). The right side includes the appropriate subsystem-level integration and testing protocols to validate that subsystem-level components are meeting the requirements.

Component level: This is the lowest level, which on the left side includes the creation of component design elements (mechanical, electrical, and software) that are inputs into the BOM creation process. On the right side, component-level testing is specified to verify that each component performs as required to meet its specification.

7.6.2.4   Disadvantages of the Traditional V-Model with Agile
Hardware and Software

As has been demonstrated, the traditional V-Model for hardware product development does bring many advantages:

- The process is predefined with tangible, logical phases and it creates an understanding of the relationships between the elements of those phases, making it easier to monitor progress and ensure as few requirements as possible are missed.
- As part of the logical steps of the work, the testing required for each phase is defined (and a test document written) before the implementation work begins and allows for testing of each element to be conducted as soon as possible to ensure compliance to the requirements at every stage.
- It ensures that design, development, and testing are given equal focus throughout the product lifecycle.
- Managing the development process becomes more transparent because of the phase approach to the work.

When adopting an agile approach to hardware development, alongside agile software methodologies, there are multiple disadvantages to using this traditional V-model:

- The step-by-step approach is extremely rigid with very little flexibility for change.
- Both hardware and software are developed during the component stage with little prior chance to experiment with prototypes or partial builds of the riskiest elements of the hardware.
- Software development cannot truly start until the first full hardware platform is created.
- The inflexibility for changes in requirements that might occur during development usually means retreating a whole stage and redefining and redeveloping the hardware.

The goal of this document is to show how to deliver working models of both hardware and software for your product. However, cost and time management are essential to produce hardware-based products that can be manufactured for profit in the market window. Therefore, you can adopt a hybrid version of this model to try to take advantage of both worlds.

Creating early models of the hardware design, using simulation or emulation, gives the agile software team ways to test its sprint deliverables before any hardware is built. Starting with hardware product "systems" made from commercially available boards allows interfaces to be tested and management software work to begin alongside firmware development – even if this "system" prototype is the size of the workbench instead of being a nice piece of custom hardware that is closer to the final product.

Iterating rapid prototypes, as well as creating partial experimental hardware pieces alongside the agile software development within the NPDI phase-gate process, is a good compromise that suits the development of hardware products. Mechanical design iterations can accompany the hardware design activities but are likely to be less frequent because the final physical design of the hardware needs to be complete before the final mechanical parts can be tested, but some rapid prototyping can be achieved in parallel.

It is never going to be possible to get a hardware "sprint" to the two weeks of a typical software agile process. The modified V-model discussed in Figure 7.17 has been redrawn in Figure 7.18 to include agile hardware product realization techniques. Because of the needs of hardware development, upfront definition is still required as well as the full integration testing stages, but in between you can see the use of both agile hardware and software iterations.

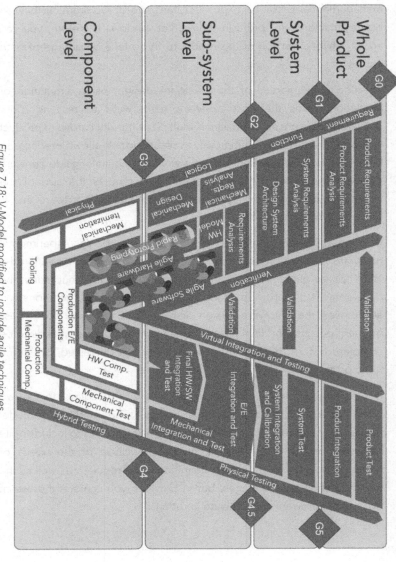

Figure 7.18: V-Model modified to include agile techniques.

7.6.3   How Does Product Testing Support the Product Realization Process?

Like many of the other best practices, manufacturing tests should be incorporated into your product development process as early as possible, from the design and planning through production phase, and should address field failures, returns, and debugging. Depending on your specific product profile, you should select the elements that best fit your needs. The principal types of testing are as follows:

- **Pre-prototype testing:**
    - Design testability into the product.
    - Create test specifications.
    - Develop a test strategy and a test roadmap.
    - Incorporate product DfT techniques.
- **Prototype testing:**
    - Identify low-cost test methodologies.
    - Incorporate as much test capability into the design as possible.
    - Leverage knowledge of the prototype test into the next level (pilot and production).
    - Provide test feedback to design and test engineers.
- **Pilot testing:**
    - Utilize flying probe testing.
    - Perform bench-top boundary scans.
    - Incorporate automatic optical inspection.
    - Use X-ray imaging for solder joint(s) inspection.
    - Develop in-circuit and functional tests where practical.

-- continued --

- **Volume production testing:**
  - Incorporate automatic optical inspection.
  - Use X-ray imaging for joint inspection.
  - Design in-circuit (board-level) tests for printed circuit boards to identify manufacturing defects.
  - Perform functional (board and system level) testing.
    - Automate the testing to minimize operator involvement where possible.
    - Verify that the product performs to customer specifications.
    - Ensure proper documentation and training are provided for production.
    - Minimize test times as testing will often be the gate in production.
  - Perform burn-in testing.
    - This test will induce potential failures in manufacturing before the product gets to the customers.
    - Such testing may be reduced or eliminated over time based upon yield data.

See Section 0 This best practice has numerous implications for the future quality perceptions of your product, so it is described in two parts: product reliability and product validation and testing.

Product Reliability for more information about reliability.

Key Insight: Product validation and testing.

Testing is an area where a strategy and investment early in the development process is critical. If electronic PCBs are designed without proper consideration for testing, a full redesign is very likely to be required to incorporate this capability, adding both time and cost to your product realization process.

7.7 Meet Agency and Environmental Compliance Requirements

Navigating through the various regulatory and environmental compliance requirements these days can feel like you are looking at an ever-changing alphabet soup of regulations. Many companies will try to minimize or ignore these requirements to speed up their time to market, only to find out that their product fails critical requirements or that their business cannot scale up, as we saw in the earlier example of 23andMe. Figure 7.19 shows some examples of compliance requirements.

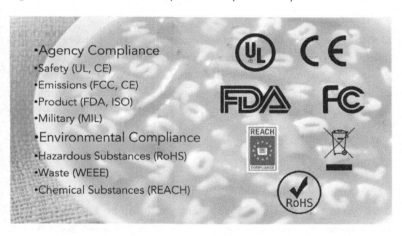

Figure 7.19: Examples of compliance requirements.

7.7.1 What Are the Main Drivers for Compliance?

When it comes to compliance, there are a variety of drivers including governmental laws, regulations, and standards (e.g., those of the FDA and FCC), industry standards (e.g., NEBS), regulatory bodies (e.g., ISO and UL), market and commercial acceptance or differentiation, and potential liability risk (product safety).

7.7.2   Compliance Testing and Enforcement

Compliance testing and enforcement will vary based upon the type of product and the regulatory requirements involved. In general, the more risk involved with the product, the more testing will be required and the more serious the consequences of an infringement.

Compliance testing ensures industry acceptance and compatibility via test labs. Such tests must be submitted directly to the regulatory agencies. Self-certification, reporting, markings, permits, and licenses are also involved. Acceptance is based upon customer and user demands.

Compliance enforcement entails Inspections by regulatory bodies and/or independent auditors. It can be triggered by competitor complaints. Standard import and export inspections are performed. Lack of compliance can result in shipment holds issued by regulatory authorities. Lawsuits based upon a product's failure to meet compliance standards can also be filed.

7.7.3   Impacts of Compliance Failure to Your Business

As illustrated in the compliance failures of Raynet and 23andMe, the most typical impact is a shipment hold, which can be fatal for an early-stage company and have significant cost and market setbacks for established companies. The impacts of compliance failures include

- delayed market launches;
- inability to ship products,
- increased costs for recalls, redesign, retesting, and logistics,
- lost market opportunities, and
- brand erosion.

7.7.4   Common Types of Compliance

You will need to meet some types of standards and regulations for your product before shipping to your customers. You should understand and be able to address the compliance landscape for your product as part of your NPDI process. Typical compliance types include:

- safety, including fire, electrical, physical injury, and Occupational Safety and Health Administration regulations;
- electrical emissions, radio, and susceptibility;
- hazardous materials;
- waste recycling;
- Environmental Protection Agency (air and water), if applicable;
- FDA, if applicable;
- military, aviation, etc.;
- supply chain and sourcing;
- compatibility and interoperability standards and requirements;
- industry commercial requirements;
- export and import regulations; and
- labeling, reporting, and documentation.

7.7.5   Elements of a Compliance Program

A strong compliance program will reduce the risk of regulatory compliance failures and potential product safety and or field recalls. You should develop a pre-submission plan early and iterate with the regulatory body as you go through the process to avoid last-minute surprises.

7.7.5.1   Creating Supporting Requirements Documents

This is where a comprehensive and complete set of MRD and PRD along with an EDS document will come in handy. They will provide you with an understanding of what types of compliance your product will need to meet as well as which countries you plan to ship to over time.

7.7.5.2   Assessing the Compliance Landscape

Achieving compliance goes more smoothly when you have a clear understanding of which standards and regulations your product needs to meet. To optimize the compliance process, you should include an expert in this area to review and assess the landscape. You may also want to look at datasheets from your competitors as a reference point. If the FDA is the regulator, you will want to review predicate devices.

Compliance requirements may vary depending on your product's specific design characteristics and the countries where you intend to ship the product. Understanding what requirements your product needs to meet is an essential part of a robust compliance plan.

7.7.5.3   Developing a Compliance Plan

The compliance plan will include the resources required, desired test laboratories, time and cost estimates, and a table listing the certifications and standards your product will need to meet over time. The plan should allocate sufficient time and resources for product "pre-scan" testing, prior to formal laboratory submission. The "pre-scan" should be supported by personnel who understand the laboratory requirements but are not associated with the selected test lab or regulatory body. It is common for test laboratories to recommend more tests than are required to gain compliance, so part of the planning process should be to clearly define which tests are required.

7.7.5.4   Implementing the Plan

An experienced project manager or compliance engineer should be assigned to identify and engage testing laboratories to navigate the testing and reporting process. As part of the implementation, the project manager coordinates with the lab for construction reviews and product testing within the plan's time and budget. A successful project needs formal approval from the testing body to label the product and meet the regulatory requirements.

7.7.6 Managing Certification Costs

It is often less expensive to plan multiple country certifications during the same test window at a certified laboratory. Figure 7.20 shows an example of costs for a 3D printer product from 2019.

Certification Type	US Only	US, Canada	Mexico	Europe Only	US, Canada, Europe	Japan	Korea
Emissions	✓	✓	✓	✓	✓	✓	✓
Safety	✓	✓	✓			✓	✓
RoHS				✓	✓	✓	✓
REACH							
WEEE				✓	✓	✓	✓

When done separately, all of the required tests could cost between $150,000 and $250,000 for each model of the product but, when tested as a complete package for all countries and standards, the total could reduce to between $40,000 and $100,000

Figure 7.20: Example of compliance certification requirements and costing.

7.7.7   Tips for Compliance Success

Here are some tips for achieving successful compliance:

1. Plan for compliance before starting product development.
2. Create a thorough product requirements document.
3. Identify into which countries the product will sell over time.
4. Define who will operate the product and in what setting (consumer, industrial, medical, etc.).
5. Create a user manual and installation and maintenance manuals if needed. Regulators judge safety based upon how the product is intended to be used and in what environment.
6. Identify which compliance and certifications are needed. Engage with a nationally recognized testing laboratory (NRTL) to evaluate the first prototype. It can refine applicable standards and provide a preview of critical areas of concern.
7. Conduct formal tests on the first stable (near-final) hardware.
8. Allow ample time (at least 8 to 16 weeks) to test and revise the hardware.
9. Document all parts and implement a change control system. Changes may require recertification by the lab.
10. Employ a dedicated program manager to coordinate and synchronize internal and external compliance activities.

> **Key Insight:** Meet agency and environmental compliance requirements.
>
> Fully research all the compliance requirements your product needs to meet in every country you want to ship it to. Add safety testing to your plans to help mitigate risk in the field.

7.8   Deploy Scalable Business Systems and Processes

When it comes to scaling a hardware product-based business, you should create a business systems strategy and plan that will keep your product data under control and enable you to smoothly scale your business, without developing isolated "silos" of data that are common in ad hoc NPDI environments. Figure 7.21 shows the standard business systems tools that companies deploy within the product lifecycle.

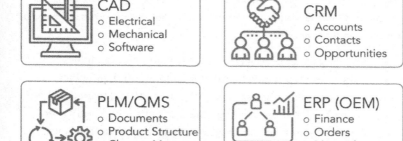

Figure 7.21: Examples of business systems.

7.8.1   Why Use Business Systems to Manage Your Product Data?

Business systems are critical for supporting agile product realization, optimizing your processes, meeting regulatory compliance, and scaling your business. Manual processes are inherently error prone and will create isolated silos of information, which will slow down your NPDI process and impede your ability to share information between groups and with suppliers. Without effective database tools, your product data will be subject to "tribal knowledge," which means that, if you lose key development resources, your data may not be recoverable. In addition, there is no way to effectively search your data in silos, so having information readily available for regulatory and/or customer audits will be difficult to impossible. Finally, strong project management and key performance indicators can be enabled with business systems, which will support a more agile and innovative NPDI methodology.

7.8.2   What Are the Business Drivers for Implementing Business Systems?

As you move your product through the product lifecycle, you will find an increasing need to shift from flexibility in the concept phase to greater control as you approach volume production, where you will have to share increasing amounts of product data with both internal and external partners. Having business systems in place that are easy to install and manage (ideally cloud based) will provide the systems infrastructure required to scale your business and support agile hardware NPDI practices. Common business drivers for implementing business systems include the following:

- Enabling faster product development and introduction.
- Supporting rapid development iterations and greater innovation.
- Improving data accuracy and integrity (reducing manual entry).
- Searching for and accessing data more effectively and faster.
- Automating manual processes and implementing workflows.
- Integrating and automating the flow of data between systems.
- Complying with ISO and FDA regulations (using electronic signoff).
- Supporting more effective project management.
- Tracking key performance indicators.
- Maximizing the use of your budget.

7.8.3 Navigating Standard Automated Business Systems for Agility and Scalability

Standard automated business systems provide a comprehensive suite of tools designed to facilitate an agile, hardware-based product NPDI process, while simultaneously promoting business scalability. These systems encompass a wide range of functionalities, from customer relationship management, which starts during the concept phase (but continues for the life of the customer but not just the product), to engineering CAD and simulation tools for initial design creation. Subsequently, product lifecycle management systems and quality management systems structure the raw product data into accurate product records, while maintaining quality controls. Lastly, the process culminates in materials procurement and product assembly, managed by enterprise resource planning systems and manufacturing execution systems. Figure 7.21 offers a visual representation of these systems, which are further detailed in the following.

7.8.3.1 Customer Relationship Management

A Customer Relationship Management (CRM) system is a software solution specifically designed to help businesses effectively manage, monitor, and optimize interactions with both existing and potential customers. By centralizing customer data, automating communication processes, and providing advanced analytical tools, a CRM system can enhance customer satisfaction, retention, and sales performance. By streamlining sales, marketing, and customer support functions, a CRM system empowers businesses to build and maintain robust, long-lasting relationships with their clientele. In an integrated systems environment, a CRM can seamlessly communicate with downstream PLM and ERP systems to ensure smooth operations.

7.8.3.2   Electrical CAD

Electrical CAD (eCAD) software systems are used to design and develop electronic systems such as PCBs and integrated circuits. eCAD software enables electrical engineers to create and modify diagrams and layouts including two- and three-dimensional models. eCAD is also known as electronic design automation. These systems create the raw parts data used in downstream activities and often feed into a PLM–QMS.

7.8.3.3   Mechanical CAD

Mechanical CAD (mCAD), also referred to as mechanical design automation, software enables mechanical engineers to create product designs with accurate and detailed technical drawings (two- or three-dimensional models) and specifications for engineering and manufacturing. These systems create the raw parts data used in downstream activities and often feed into a PLM–QMS.

7.8.3.4   Software Management Systems

Embedded software and firmware tools include platforms for the development of software that is embedded into hardware systems using standard operating systems, languages, and tools. These systems create the raw software programs used in downstream activities and often feed into a PLM–QMS.

7.8.3.5   PLM

A PLM system is an enterprise-level system that manages all product records, simplifies the design-to-manufacturing process, enables cross-functional collaboration, and supports regulatory compliance in all phases of the product's lifecycle. It is an important tool for organizing raw parts data from the engineering world (e.g., eCAD, mCAD, and software) into a complete product record that can then be communicated to other internal systems, such as Enterprise Resource Planning (ERP) and customer relationship management, as well as outside vendors, such as contract manufacturers.

7.8.3.6   QMS

A QMS is a formalized system that documents processes, procedures, and responsibilities for achieving quality policies and objectives. A QMS helps coordinate and direct an organization's activities to meet customer and regulatory requirements and improve its effectiveness and efficiency on a continuous basis. QMS functionality is commonly (but not always) linked as a module to a PLM system. In Figure 7.21 and Figure 7.22 we combine PLM–QMS into the same graphic to simplify the illustration.

7.8.3.7   ERP

ERP systems manage the manufacturing, logistics, distribution, inventory, shipping, invoicing, and accounting. ERP software can aid in the control of many business activities, such as sales, marketing, delivery, billing, production, inventory control, quality management, and human resource management. These systems will take product information from a PLM system (or from CAD systems if a PLM is not available) and add information (e.g., part cost, lead time, and lifecycle) that is required to support the materials management process.

7.8.3.8   Manufacturing Execution System

A Manufacturing Execution System (MES) is a computerized system used in manufacturing to track and document the transformation of raw materials to finished goods. An MES provides information that helps manufacturing decision-makers understand how current conditions on the plant floor can be optimized to improve production output. An MES works as a real-time monitoring system to enable the control of multiple elements of the production process (e.g., inputs, personnel, machines, and support services). The MES is commonly owned by the CM and interacts primarily with the CM's ERP system.

7.8.3.9   How Do Data Flow through the Product Lifecycle?

As discussed above, data flow through the various business systems in a similar manner to flow in the development process. Figure 7.22 illustrates the typical flow of data between systems in an OEM through communication with a CM. A strong business systems strategy will automate the exchange of data between systems and keep primary control of each data element in the system for which it was originally created.

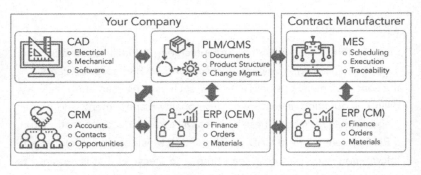

Figure 7.22: Example of data flows between business systems.

7.8.4   Streamlining Product Data Flow across Business Systems

Once you establish a robust systems infrastructure, it is common for many product data elements to be duplicated across different business systems. Take part numbers as an example; they are generated in engineering, but they are also utilized in PLM, ERP, CRM, and MES systems. To adhere to good database management practices, unique data elements should be controlled within a single system, even if they are replicated in others. When properly architected, data can be automatically duplicated into other systems through a seamless, automated, and integrated process. Identifying which systems serve as the "data master" (originating and controlling data) and which function as the "data copy" (duplicating data from the data master) is critical to your data management strategy. Figure 7.23 provides a partial real-world customer example of how this mapping can be accomplished.

System	Item/Part Masters	Production Cost	Vendor/Supplier Master	Purchasing	Inventory	Master Production Schedule	Financials	Order Fulfilment	Price List	Customer Master	Sales Forecast/Pipeline	Sales Orders/Shipments	Manufacturing History	Ship History	RMAs
Automated Data Links Example															
Your Company															
CRM	C	C						C	C	C	M	C		C	M
ERP	C	M	M	M	C	M	M	M	M	M	C	M		C	C
PLM/QMS	M	C	C												
Contract Manufacturer															
ERP	C	C			M	C								M	
MES													M		

M   Data Master

C   Data Copy

Figure 7.23: Example of data mapping across business systems.

7.8.5   Tips for Successful Business Systems Deployment

Now that you have a better understanding of the importance of business systems throughout the NPDI process, it's essential to follow these key recommendations to ensure a successful business system deployment:

- Develop an end-to-end systems strategy, considering outsourced partners and data flows between systems.
- Secure management support before investing time and resources.
- Utilize Software as a Service (SaaS) platforms, when possible, for scalability and ease of management.
- Allocate time and budget for process definition, training, and automation of data exchange to improve efficiency.
- Appoint a dedicated project manager and provide ongoing support resources.
- Establish clear Statements of Work (SOWs) and create service-level agreements with software vendors.
- Ensure ample end-user training and reserve a budget for vendor support and systems update management.

> **Key Insight:** Deploy scalable business systems.
>
> A good business systems strategy will support agile development early and volume manufacturing as you scale. Before implementing a business systems strategy, you should have developed manual or semiautomated processes based upon best practices that can be automated and scaled by leveraging business systems.

7.9   Develop a Resilient Supply Chain

Creating a resilient supply chain will allow you to handle changes in manufacturing location and availability of parts and subassemblies with little impact on your product shipments. Figure 7.24 shows the typical elements of such a supply chain system.

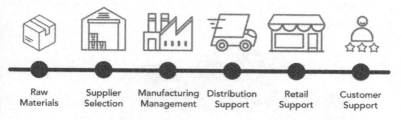

| Raw Materials | Supplier Selection | Manufacturing Management | Distribution Support | Retail Support | Customer Support |

Figure 7.24: Typical elements of a resilient supply chain.

Once you enter volume manufacturing, the cost of materials makes up the largest percentage of cost for most high-tech products. Making sure that the right parts are designed into your product early will help you meet your cost, quality, and delivery target as your business scales.

You will also need to manage your cash flow, given that you will need to procure parts, assemblies, and test products; ship to distribution sites; and, finally, sell to customers. In most cases, you will not receive revenue until the very end of the process – after customer payment. To maximize cash flow, you will need to manage the order fulfillment process carefully to avoid cash flow being delayed or wasted.

When transitioning from prototype to volume production, you should understand the costs and allocation of resources associated with building in lower volumes versus higher volumes. We will look at some characteristics of both prototype production and volume production for electronic hardware products in the next sections.

7.9.1   Elements That Make Up Total Product Cost

As you shift from engineering development into volume production, there will be a corresponding shift in the elements that make up your product cost. The cost for making a few units in the prototype phase may vary dramatically from the cost for making these same units in volume production. Figure 7.25 shows common elements that make up the total product cost.

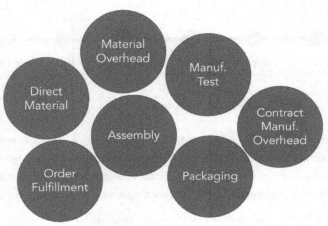

Figure 7.25: Elements that make up total product cost.

Figure 7.26 represents the relative cost proportions of each element when these elements are mapped into a prototype build and then a volume product build. You can see that, as a proportion of the total cost of a single prototype or volume product, different elements have a greater or lesser effect on the overall cost – as shown by the relative sizes of the element circles. The relative cost difference between a prototype and volume production unit is also represented by the size of the outer circle. Paying attention to the best practices discussed in this book will allow you to focus on getting these proportions correct at the same time as controlling the overall cost of the finished volume product.

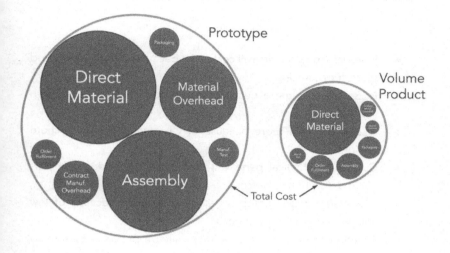

Figure 7.26: Proportions of total cost between prototype and volume.

7.9.2   Characteristics of a Prototype Build (U.S. Manufacturing)

While the elements associated with a prototype build will be like that of a production build, the prototype costs will be substantially higher because of lower quantities as well as higher pricing for rapid turnaround and onshore assembly pricing.

Expect a wide increase in costs from 150% to more than 500% in moving from prototype to volume production, depending on the product profile and parts involved. The characteristics of a prototype build are as follows:

- **Electrical components will cost more** because of low quantity and quick-turn sourcing.
- **PCBs will cost more** because of low quantities and quick-turn sourcing.
- **PCBAs will cost more** because of low quantities and onshore pricing.
- **Custom mechanical parts will cost more** because of the low-volume 3D printing process.
- **Box assembly** prototype units will **cost more** because of low volume and onshore CM rates.
- **Functional** prototype tests will typically be performed internally by the OEM and "expensed" using engineering development resources.
- **CM on-hand add-ons** will be more numerous because of local CM rates.
- **Pack-out, packaging, and order fulfillment will cost less** because these are not typically needed to support prototype builds.

7.9.4   Characteristics of a Volume Product Build (Offshore Manufacturing)

As you make the shift from prototype into volume production, for most products the elements of cost will shrink substantially **because of** volume pricing of materials, the use of production tooling, and efficiencies gained from automated production processes. The characteristics of volume production build are as follows:

- **Electrical components** will cost less using standard distribution channels for parts.
- **PCBs** will cost less because of volume offshore pricing.
- **PCBAs** will cost less because of standard lead times and offshore rates.
- **Custom mechanical parts** will cost less because of production tooling at the offshore vendor.
- **Box assemblies** will cost less because of higher volume and offshore rates.
- **Functional testing** will cost less because of production volumes and offshore rates.
- **Pack-out, packaging, and order fulfillment** will cost more.
- **CM overhead add-ons** will cost less because of offshore production rates.
- **Transportation and logistics** will cost more to support offshore manufacturing.
- The **cost of goods sold** will be lower because of volume pricing and optimized assembly processes.

7.9.5   Order Fulfillment and Logistics

Getting products from your factory to the end customer is an often overlooked and underestimated activity that should be included in your product realization strategy and plans.

7.9.5.1   Order Fulfillment

Your order fulfillment strategy varies based on the type of product you are developing, the manufacturing volumes, and the end buyer for your product. For instance, low-volume, complex business-to-business products typically ship directly from your factory or CM to the end customer. Alternately, high-volume business-to-consumer products commonly engage a 3PL provider to support fulfillment.

Figure 7.27 illustrates a typical process flow for order fulfillment.

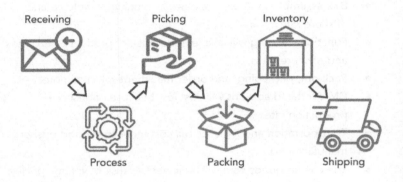

Figure 7.27: Order fulfillment flow.

If you have multiple subassemblies built overseas, you may want to consider engaging a fulfillment house that can support final assembly and can test, label, and ship to optimize your supply-chain strategy. Consumer and Internet of Things devices are usually shipped in bulk to the fulfillment house that can perform the final configuration and packaging.

7.9.5.2   Packaging and Labeling

Packaging and labeling are important to the look and feel of your product, as well as for meeting regulatory compliance requirements (e.g., UL, FCC, and FDA). Where possible, you should include bar codes to improve product traceability in the factory and in the field. Packaging and labeling design should be included as part of your NPDI schedule and concurrent engineering gate review process.

7.9.5.3   Returns Management

Your responsibility for the product does not end upon customer receipt. Don't forget to include returns management as part of your product realization strategy and NPDI plan. A typical process flow for this is shown in Figure 7.28: Typical customer journey including returns flow.

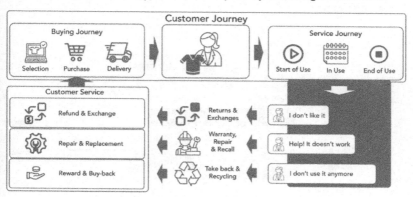

Figure 7.28: Typical customer journey including returns flow.

There are several returns management strategies that should be considered for an effective returns program. Strategies for returns range from return and replacement for simple, low-cost products all the way to sending a skilled technician onsite with spare parts for complex products that are difficult to move. Here are some considerations that may influence your strategy:

- Product complexity and cost: Can the product be easily packaged and shipped at a reasonable cost?
- Location: Where are products being returned from?
- CM selection: Can your CM support returns and repair?
- 3PL selection: Can your 3PL support returns and repair?

- Quality management: How are you going to collect root-cause failure information?
- Quality systems: Does your systems strategy support field failure data collection and analysis?
- Customer service: What is your customer communications and replacement plan?

7.9.5.4 Logistics Management

Logistics is often a "hidden" cost that should be considered as part of your total cost (as highlighted as an element of "Accounting for "Total Cost of Ownership" above). Logistics costs are manageable with good planning. A clear logistics management strategy will help minimize these sneaky expenses. Figure 7.29 shows the main elements of logistics management.

Figure 7.29: Typical logistics management cycle.

7.9.6   Tips for Developing a Resilient Supply Chain

Greater uncertainties in the global economy are rapidly increasing the risk to supply chains. These include aspects such as environmental disasters and pandemics, geopolitical instabilities, and materials shortages. While it is impossible to eliminate risk altogether, there are several strategies you can implement to increase your supply chain resilience. Some examples of these include the following:

1.   Understand the end-user market location, capabilities, politics, and tariffs.
2.   Utilize standard and short-lead-time parts in design.
3.   Stabilize your product design prior to scaling into volume.
4.   Select suppliers with a good fit with your product for technology, culture, volumes, support, and cost. Also consider the supplier's financial stability, size, and ability to scale to your needs.
5.   Duplicate supply sources where practical.
6.   Communicate clear and predictable production volumes.
7.   Hold regular supplier performance reviews.

Key Insight: Develop a resilient supply chain.

Think of your key suppliers as an extension of your business – they need to be effectively managed with mutual understanding of expectations, clear communications, and active management over time including key performance metrics such as quality, cost, and delivery.

7.10   Verify Readiness for Volume Manufacturing

Costs for mistakes can escalate rapidly when moving from prototype to pilot to production. Therefore, resolving design problems and optimizing product manufacturability by implementing the best practices listed above and actively addressing the issues found as a result of feasibility, prototype, and pilot builds will help prepare your product for volume manufacturing. The following paragraphs provide some critical questions to answer to confirm you are ready for volume manufacturing.

If you have made it successfully through the product development process and are ready to make the leap into volume manufacturing, congratulations on this accomplishment! Yet, before you proceed, take a look at the following questions to make sure you have all your bases covered:

- Have lessons learned from feasibility and prototype builds been taken and incorporated into the latest product iteration?
- Have all critical DfX requirements been met?
- Has the required regulatory certification been achieved?
- Do you have a volume materials sourcing strategy in place?
- Are prototype and production tooling ready and available in production?
- Do you have production test plans and fixtures available in production?
- Is final packaging defined and available in production?
- Is the product properly documented and is change management in place?
- Do you have manufacturing documentation and training ready for assemblers?
- Are distribution, spares, and servicing plans in place?

Depending on your unique product profile and/or business dynamics, there may be other items you will want to include, but this list is a good starting point

Key Insight: Be ready for volume manufacturing.

This is where you bridge the chasm between prototype and volume manufacturing. Make sure you have your documentation complete and located in automated systems that will enable you to clearly communicate product data as well as manage changes as you scale into volume production. Tribal knowledge does not work here!

8   Managing the Agile Hardware Lifecycle

In the previous sections of this book, we've covered the core principles of agile hardware product realization, the importance of understanding your markets, highlighted the differences between agile hardware and software, shared strategies for implementing agile hardware, and introduced the 10 best practices. Now, we're ready to put theory into practice by offering a management system that unifies these elements together, creating a progressive flow of activities that accelerates your product from concept to volume production.

There are several parallel activities that offer practical ways to drive progress, ensure visibility of goals, and provide indications of success that will give you confidence in meeting your product launch timelines and help you gain satisfied customers. These activities support the principles of agile hardware product realization and the 10 best practices, and include the creation of product documentation, applying effective project management techniques and measuring success with key performance indicators (KPIs).

8.1   Documenting Your Product

Throughout this book, we've deliberately kept the topic of documentation to a minimum. This chapter, however, aims to shed light on the "how" and "what" of documentation, because documentation is what facilitates the transition from experimenting in the lab to building products at scale, without the need for engineering oversight.

There are many advantages to properly documenting your product from concept through design, development, and validation. The primary benefit is that your entire team, including your CM and outside partners, can refer to the most recent and accurate product record, as the product makes its journey through the product lifecycle. In addition, accurate and complete documentation is necessary to support outside regulatory bodies (like UL, FCC, and FDA), field support and end customers.

It is worth noting that some companies may choose to create documentation later in the lifecycle, prior to engaging a CM. However, our experience suggests that this often leads to errors and omissions of critical information that results in schedule delays, quality defects and increase costs. Some common examples include failing regulatory compliance due to incomplete documentation, buying the wrong parts thanks to inaccurate documentation, and installing down revision firmware in production because of uncontrolled documentation.

While some companies may choose different naming conventions than those provided in this book, the documentation list shared here offers a view of the standard documentation required for a hardware-based product company to successfully navigate the product lifecycle.

Product documentation starts when the requirements are described in the MRD, which is then broken down into phased feature releases in multiple PRDs. These documents and their impact on the product development cycle were also described in section 6.2, but we're including them here for completeness.

The PRD describes functionality for an individual product release, which makes up a defined subset of features specified in the MRD. This documentation is then used to create the detailed design requirements in the (multiple) EDS(s) (e.g., electrical, mechanical, software). Figure 8.1 below illustrates how the high-level requirements of the MRD and PRD are mapped to the lower-level specifications in the EDS(s).

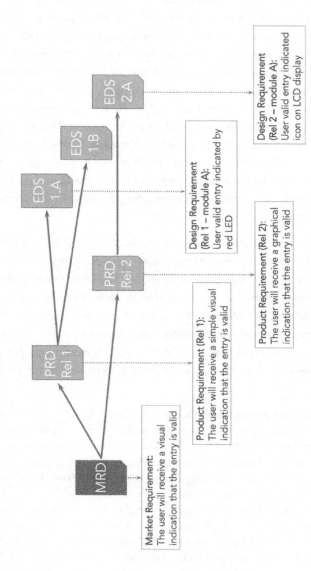

Figure 8.1: Product requirements hierarchy example

As shown in Figure 8.1, there is a high-level requirement for the user to receive a visual indication that they have entered something correctly. This requirement is then translated across two product design iterations with increasing refinement. In the first release, this requirement is deliberately simplified: a single LED illuminates as the indicator. This simplified solution fast-tracks the product's market entry, while fulfilling the market's demand for visual feedback. In the second release, the product requirement generates more complex design requirements. This increased complexity delivers higher product performance for the customer of a visual indicator to achieve the same goal. The result is that a simpler and faster design is enabled in release 1, which speeds the product to market, and the product's performance is improved in release 2, which delivers better product performance for the customer. In both cases the market requirement of providing a visual indication to the customer is met.

To continue the product journey, we highlight below the key product documents, their accompanying validation documents, and their subsequent reports. The documents should align with the PRD (for the product release) and the EDS(s) (for the detailed design) as well as the testing description needed for production test. Typically, these documents, and their revision history, are managed in a PLM system, ensuring a comprehensive view of all relevant documents for each product release. The list below provides a foundational list of these documents:

- **Market Requirement Document (MRD):** A high-level document that describes the full scope of the product over its lifecycle based on current market knowledge and requirements. As additional market demands surface, and as the product gains market traction leading customers to ask for new features, this document will be updated.
- **Product Requirement Document (PRD):** A high-level document describing a single product release based on achievable features extracted from the MRD that can be incorporated into a market-ready product. There will normally be several PRDs developed throughout the product lifecycle, each describing a distinct evolution of the product's feature set, starting with the first product shipped to customers.

- **Electronic Design Specification (EDS):** Usually several detailed documents which map the features required in the PRD to designable product elements, such as PCBs containing hardware functionality or interconnecting sub-systems. These documents describe a low-level block diagram of functionality, indicating which features will be delivered by combinations of hardware, firmware, and software (as necessary) from each logical block.
- **System Test Specification (STS):** A high-level validation requirements document that describes how the complete product release (as per the relevant PRD) should perform. This represents a set of high-level validations proving PRD requirements.
- **System Test Report (STR):** A document that records the results of the validations carried out against the STS. Because these validations may be repeated to fix bugs or failures, a new test report is created for each iteration of the validation suite.
- **Customer Trials Specification (CTS):** A document describing the expected customer-validation performance results of the controlled introduction product examples. It includes details like what will be provided to the customer, how many customers need to be signed up for this trial, what is expected of these customers, how you will record the results, how you will handle failures, and your support organization, during the customer trials.
- **Customer Trials Report (CTR):** This document records the results of the customer trials, including observations from your team and the customers involved. This report will be used to inform the final sign-off for volume manufacturing and will usually provide guidance for new features (missing per the customer) as well as how to manage discovered failures.
- **Design Test Specifications (DTS):** These documents describe how the functional blocks specified in the EDS will be validated to ensure they perform as required. This document should also include descriptions of standard test parameters, waveforms and/or data, so that testing can be repeated with known inputs effectively across multiple versions of the product during agile iterations.
- **Design Test Report (DTR):** This document provides evidence that all the validation tests required showing functional blocks perform adequately were done and records the results obtained during those tests. Failures and suggested fixes will

be noted in this document, and multiple versions of this document may be required as agile hardware iterations are performed.

- **Manufacturing Documentation Package:** A complete set of product-release-coherent documents which tell the CM what is needed to build the product including:
  - o Bill of Materials (BOM)
  - o Mechanical Drawings
  - o PC-board Layout Files
  - o Assembly Drawings
  - o High-resolution Artwork
  - o Quality Documents and Procedures
  - o Test Procedures with Pass/Fail Criteria
  - o Final Test Criteria

We have intentionally left out documentation around the business process, such as the quality manual, control of documents, and training as these documents are already well covered by ISO guidance and many other books.

Figure 8.2 below shows when these documents are produced throughout the product lifecycle.

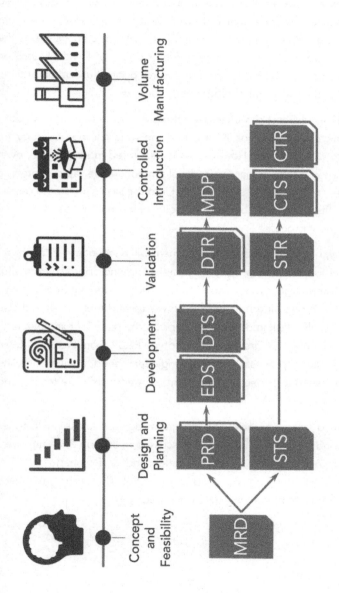

Figure 8.2: Diagram of documents in lifecycle

8.2   Project Management

In section 5, we highlighted the significance of creating a comprehensive plan that includes all the people and activities involved in developing your product. It is important to have a clear plan for how you will get from concept to volume manufacturing This plan should delineate the different stages of activity, internal interactions, and critical decision points along the way. By doing so, it ensures optimal utilization of time, resources, and personnel throughout the process.

Traditionally, a Gantt chart has been the preferred tool for creating this plan, which has been closely tied with the waterfall methodology for hardware development. However, as we explored in section 7.6, even with the constraints of hardware development, agile rapid prototype iterations can be seamlessly integrated into the plan prior to the design freeze. This brings the benefits of agile practices into the hardware development process.

Inevitably, hardware product development will encounter critical phase-gate decision milestones, typically centered around approval points for further, often larger, investment. Your project plan should cover these milestones, monitoring high-level progress toward them. It should also include detailed task information to support the plan, for example, long lead-time component ordering or prototype PCBA manufacturing. Even in a fully agile process, software engineers need a clear timeline indicating when to expect a physical prototype for software testing and refinement.

Maintaining clear visibility of progress against these milestones (corresponding to the Phase Gates discussed in section 2.3) empowers you and your company to better understand and manage the risks associated with bringing your product to market. This ensures that each investment made along the way is optimized.

8.3   Managing Change

Change is inevitable, and this is especially true in the realm of hardware product realization. Your product will invariably evolve over time, errors and omissions will be discovered, failures will occur which require fixing and hardware components may become scarce or obsolete. Managing such changes requires a systemic approach. That's where change control comes into play. This methodology allows you to collect, track and analyze changes to your product and processes, ensuring you learn from mistakes while keeping all the product information current, accurate and controlled.

Establishing a change control process throughout your product lifecycle enables you to keep your product record up to date, inform your engineers, CM, partners and customers of issues and their solutions, and maintain the quality of your product throughout its lifecycle. Regular reviews of changes should be scheduled, supplemented by high-priority ad hoc meetings for urgent matters that can't wait for the regular meeting (such as critical defects or bugs, safety issues or unanticipated parts shortages).

A well-structured change control process ensures that fixes are prioritized appropriately, features are added only when necessary, and volume production proceeds without interruption. We recommend the use of cloud-based electronic data systems for change control as these systems are faster to install and require less maintenance. Most PLM software packages offer this feature as standard. However, this system requires a dedicated person (not necessarily full-time) responsible for its management and maintenance. This team-member's role includes bringing changes for review and approval or rejection. This is generally performed by a change control board, which consists of senior members of the cross-functional team from marketing, sales, engineering, and CM. This individual is also responsible for convening urgent high-priority meetings for changes that need immediate review and action.

8.4 Key Performance Indicators: Measuring and Monitoring Success

In the competitive world of agile hardware product development, effective performance measurement is crucial for evaluating the success of your NPDI process and providing critical insights for continuous improvement. This section highlights Key Performance Indicators (KPIs) in the context of the methodologies and best practices explored throughout this book. KPIs are quantifiable metrics that enable teams to monitor progress, identify opportunities, and make data-driven decisions to enhance the product realization process.

We will share the most relevant KPIs for the hardware NPDI process, such as time-to-market, customer satisfaction, product quality, and return on investment, among others. To optimize your product realization process and succeed in the market, it is critical to understand and consistently track relevant KPIs for your organization. However, remember that consistently tracking a few key KPIs over time and taking appropriate actions based on these is more important than attempting to track all relevant indicators ad-hoc. The following are some key KPIs to consider:

Management Top-Level:

- **Return on investment (ROI):** This KPI measures a new product's profitability by comparing the revenue generated to the costs incurred during development and launch. A higher ROI signifies a more successful and financially viable product development process. To calculate ROI, use the formula: ROI = (Net Profit / Total Investment) * 100. Benchmark your product's ROI and other financial metrics against those of similar products in the market or within your company to evaluate performance and identify areas for improvement.
- **Customer satisfaction:** This KPI assesses customer satisfaction with a new product. Elevated customer satisfaction can foster customer loyalty, encourage positive word-of-mouth, and boost sales. Utilize surveys, net promoter score (NPS), user reviews, social media monitoring, customer support feedback, customer satisfaction index (CSI), product usage data, and sales data to obtain feedback.

NPDI Team Level:

- **Time-to-market (TTM):** This KPI measures the duration from a product's conception to its successful market introduction. A shorter TTM indicates a more efficient development process and enables rapid capitalization on market opportunities. To calculate the TTM, add the durations of each stage in the phase-gate process, from concept and feasibility to controlled introduction, while analyzing any delays to determine root causes for future improvement.

- **Market share:** This KPI quantifies the portion of the market captured by the new product. A larger market share is a strong indicator of a successful product launch and effective marketing strategies. To calculate market share, use the formula: Market Share (%) = (Your Product Sales / Total Market Sales) * 100.

- **Development cost:** This KPI encompasses the total cost of developing a new product, including resources, labor, and materials. Controlling development costs contributes to overall profitability.

- **Quality performance:** This KPI evaluates the number of defects or failures in a new product relative to the total number of units produced. A lower defect rate indicates higher product quality and a more effective development process. Utilize metrics such as uptime, failure rates, and maintenance requirements, and track warranty claims and returns to identify potential quality issues and implement corrective actions.

- **Unit cost and gross margin:** This KPI measures the cost to produce a single unit of the new product and the gross margin made from its sale. The unit cost should be low enough to allow for a healthy gross margin, ensuring the product's profitability. A lower unit cost or a higher gross margin indicates better operational efficiency and pricing strategy. This is particularly important to keep track of as you scale production, as economies of scale should, in theory, reduce unit costs over time.

- **Engineering Change Management:** This KPI tracks the number of Engineering Change Orders (ECOs) after design freeze, along with the reasons for the changes (e.g., design fixes, new product releases, supply chain adjustments, cost reductions). A lower number of ECOs indicates a stronger product realization process.

- **Cross-functional team performance:** This KPI assesses the efficiency and effectiveness of cross-functional teams involved in the NPDI process. A high-performing team can accelerate development cycles, improve decision-making, and ultimately lead to a more successful product launch. Monitor team progress against predefined project milestones and deadlines, as well as the quality of deliverables and resource utilization.

Tracking and monitoring all these KPIs on a single dashboard gives you an overview of progress in many dimensions of the product's progress against your planned dates, costs, and other targets. Figure 8.3 shows an example NPDI Team KPI dashboard.

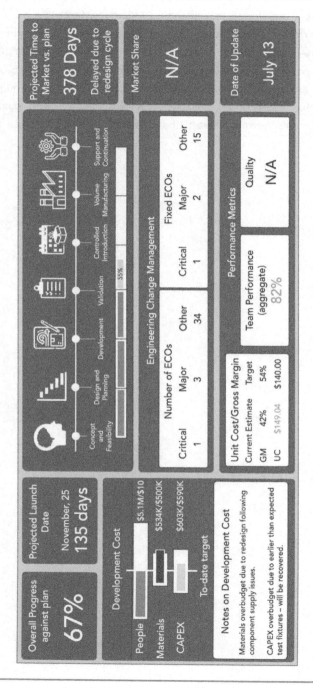

Figure 8.3: Example KPI dashboard

In conclusion, monitoring and analyzing KPIs consistently over time is essential for driving continuous improvement and ensuring the success of hardware-based product development processes. By understanding these key performance indicators, you can make informed decisions, optimize resources, and ultimately deliver innovative, high-quality products that meet or exceed customer expectations, and gain a sustainable competitive advantage in the market.

9   What Success Looks Like

9.1   Startup Company – Freewire Technologies

The success story of this startup company is described as follows:

Background:

Freewire had an extremely high demand for its initial breakthrough product, a portable Electric Vehicle (EV) charging station. It had orders before development had been completed and needed to accelerate to full-scale manufacturing to not only fulfill these but also to satisfy investors in preparation for a funding round.

The company was starting development of a second-generation product in parallel and at risk from under-resourcing, lack of regulatory knowledge, ill-defined processes, reliability issues, lack of production documentation, and supply chain issues.

Methodology:

Using the well-established practices outlined in this document, Freewire and the PRG team started by creating transparency and clarity across the entire company by

- using a defined project plan with clear dependencies,
- developing a set of test plans that could show how product specifications could be met,
- completely aligning the product and engineering teams' priorities through documentation, and
- implemented a full NPI process to scale production.

Next, because the market was immature and the product new, the challenges of a highly optimistic shipping milestone demanded action in several areas. These included

- enhancing the supply chain to ensure that all components were available when needed, and
- controlling costs to validate profitability and maximize future product potential, and

- managing all efforts within the newly established NPI process to maintain communication and solve problems dynamically across the team.

The Result:

- The development and NPI and manufacturing transfer were completely ready for the shipping milestone, and production ramped up with a few months, enabling fulfillment of orders to major companies.
- Consistent processes reduced risks to the second-generation product development, enabling faster time to market with a high-quality product.
- The resulting cost reductions and improvements in overall product quality and function led to multiple successful funding rounds as well as consistent scalable manufacturing of the product in large quantities.

9.2   Established Company – Leading EDA Business, USA

The success story of this established major company is described as follows:

Background:

This major player in the Electronic Design Automation (EDA) and simulation space had acquired a hardware simulator design company with strong Research and Development (R&D) skillsets, a lack of a formal process, but a product attractive to its market. Unfortunately, the company discovered severe problems with the uptime of the product, its performance, and its ability to repair and rework for failures.

The major player's largest customer had placed the product on ship-hold until improved reliability could be demonstrated, and the uptime of the system could be drastically improved.

Methodology:

Initially focusing on the risks to the product success and the impacts on the established business, the company together with the PRG team began by

- creating a complete project governance structure using an NPI phase-gate process,
- instituting a formal product development process including detailed documentation and checklists, and
- establishing fully integrated project plans across all divisions and departments.

Once this framework for measurement of positive change was in place, the focus turned to improving the quality of the product in question by

- establishing a reliability benchmark, plan, and desired reliability future state for the product, its subsystems, and components and
- creating and executing Hardware Accelerated Life Test (HALT) processes for critical elements of the product.

These changes closed the biggest reliability gaps in the overall system design to provide learning for newer generations of the product.

The Result:

- The product's system performance, uptime, and reliability were improved dramatically (e.g., uptime was boosted by a factor of 5 over that attained previously).
- A robust NPI and quality systems process was built around the company's needs.
- Its biggest customer removed the ship-hold and started purchasing multiple product systems.
- The improved success of this product led to greater demand for subsequent generations and increased the customer base as a result.

10 Putting It All Together

We covered a lot of territory in these nine chapters at a fairly high level with the intent to demystify the agile hardware product realization process and increase your awareness of what it really takes to get a hardware-based product to market as well as to share real-world examples and best practices. The benefits of implementing an agile hardware NPDI process and the 10 best practices outlined in this book are clear. Now, the question is "Are you ready to take action and make a real impact on your business?"

As a technical professional, you understand the importance of careful planning, execution, and continuous improvement. You also know that effective product realization requires a collaborative effort supported by a multidisciplinary team working toward a common goal. Simply cherry-picking a few tips and tricks here and there from the 10 best practices won't be enough to realize the full potential of these practices. That's why it's critical to approach the implementation of the methodologies described here holistically, with active senior management support, perseverance, a willingness to change the status quo, cross-functional collaboration, and the use of integrated business systems and robust processes. By doing so, you'll not only be able to bring more innovative products to market at scale but also avoid common pitfalls, reduce execution risk, and lower costs. You'll also experience the personal satisfaction of knowing that you've played a key role in driving the success of your business.

In the end, the choice is yours. Will you be content with maintaining the status quo and falling behind the competition? Or will you take the leap and make a real commitment to implementing an agile hardware product realization methodology with the 10 best practices of product realization all inclusively? The time to act is now. The rewards are huge for those who are willing to take the necessary steps to transform their business and bring better products to market faster.

11 Are You Ready for Scale?

You may wonder, how does my business stack up to best practices? To help answer this question, Figure 11.1 shows five key areas that you should think about to benchmark your business against to better understand how your company stacks up with agile product realization best practices.

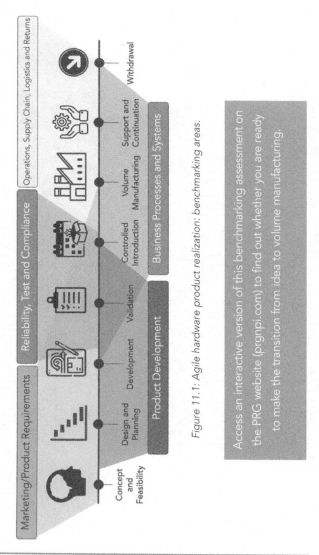

Figure 11.1: Agile hardware product realization: benchmarking areas.

Access an interactive version of this benchmarking assessment on the PRG website (prgnpi.com) to find out whether you are ready to make the transition from idea to volume manufacturing.

Index